Wayne Biddle

A Field Guide to Germs

Wayne Biddle is the author of four other books: *Barons of the Sky,* a history of the American aerospace weapons business; *A Field Guide to the Invisible,* a catalog of things beyond our senses; *Coming to Terms,* about scientific jargon; and *Dark Side of the Moon,* a history of Wernher von Braun, the Third Reich, and the space race. He is on the faculty of the Writing Seminars at Johns Hopkins University and lives in rural Maryland.

A
Field
Guide
to Germs

A
Field
Guide
to Germs

.

Wayne Biddle

Anchor Books
A Division of Random House, Inc.
New York

THIRD ANCHOR BOOKS EDITION, MARCH 2010

Copyright © 1995, 2002, 2010 by Wayne Biddle

Library of Congress Cataloging-in-Publication Data is on file.

ISBN: 978-1-4000-3051-4

www.anchorbooks.com

Nous sommes démocratisés et syphilisés.
—Baudelaire

· Contents ·

List of Illustrations   XV

Introduction   3

adenoviruses   21
anthrax (*Bacillus anthracis*)   23
arboviruses   26
arenaviruses   27
bites   28
Bordetella pertussis (whooping cough)   29
Borrelia burgdorferi (Lyme disease)   30
Brucella   34
Campylobacter   35
Candida albicans   36
chikungunya/o'nyong-nyong viruses   37
Chinese restaurant syndrome   38
Chlamydia   39

cholera (*Vibrio cholerae*)   41

Clostridium   48

**Congo-Crimean/Rift Valley/
    California/St. Louis viruses**   53

coronaviruses   55

Cryptosporidium   56

cytomegaloviruses   57

dengue   58

diphtheria (*Corynebacterium diphtheriae*)   61

dysentery   63

E. coli   65

Ehrlichia   67

Enterococcus   68

enteroviruses   68

Epstein-Barr virus   71

Fort Bragg (*Leptospira*)   73

Giardia lamblia   74

hantaviruses   75

Helicobacter pylori   77

Hemophilus   78

hepatitis   78

herpes   82

HIV (human immunodeficiency virus)   83

influenza (flu)   90

Legionella   96

Listeria monocytogenes   98

mad cow   98

Marburg/Ebola viruses   100

measles	103
mumps	108
mycobacteria	110
Neisseria gonorrhoeae (gonorrhea)	111
Neisseria meningitidis	114
papillomaviruses	116
parainfluenza viruses/respiratory syncytial virus (RSV)	116
Plasmodium (malaria)	118
pneumococcus	124
polio	125
Q fever	129
rabies	130
rhinoviruses	133
Rickettsia	134
Rocky Mountain spotted fever	136
rotavirus/Norwalk	138
Salmonella	139
San Joaquin fever (*Coccidioides immitis*)	141
slapped cheek disease (parvovirus B19)	141
smallpox	142
Staphylococcus	150
Streptococcus	153
syphilis (*Treponema pallidum*)	157
tinea	162
Trichinella	163
tuberculosis	163
typhoid (*Salmonella typhi*)	165

typhus   172

varicella–zoster virus   174

West Nile virus   175

worms   176

XMRV   179

yellow fever   180

Yersinia pestis (plague)   184

Zika fever   191

Notes to the Introduction   193

Selected Bibliography   197

Index   201

Acknowledgments   217

· *Illustrations* ·

1. An Egyptian stele, or stone tablet, from the eighteenth dynasty. *Courtesy of March of Dimes Birth Defects Foundation.*

2. French chemist and microbiologist Louis Pasteur (1822–1895) with his granddaughter. *Courtesy of National Library of Medicine.*

3. German bacteriologist Robert Koch (1843–1910) and his wife. *Courtesy of National Library of Medicine.*

4. Fresh AIDS virus budding from the membranes of infected T-4 lymphocytes. *Courtesy of Centers for Disease Control and Prevention.*

5. Young suitor kneeling before Death disguised as a beautiful girl. *Courtesy of National Library of Medicine.*

6. A woodcut from Thomas Bartholin's *Historiarium anatomicarum* (1654–1661) showing a doctor garbed against the Plague. *Courtesy of National Library of Medicine.*

7. An early-nineteenth-century caricature of hypochondria and pseudoprotective measures against cholera. *Courtesy of National Library of Medicine.*

8. Trying to make the best of quarantine on Ellis Island, New York. *Courtesy of National Library of Medicine.*

9. Science sleeps on a New York dock as Cholera floats across the ocean from Europe. *Courtesy of Library of Congress.*

10. "She May Look Clean—But." *Courtesy of National Library of Medicine.*

11. Anthrax cells. *Courtesy of Dr. Edward J. Bottone.*

12. Lyme disease poster. *Courtesy of National Library of Medicine.*

13. Cholera poster. *Courtesy of National Library of Medicine.*

14. An early-nineteenth-century German illustration of a man wearing protective gear against cholera. *Courtesy of National Library of Medicine.*

15. An illustration published in London in 1832 showing two physicians bundling up a cholera patient. *Courtesy of National Library of Medicine.*

16. Epidemic in Marseilles, 1865. *Courtesy of National Library of Medicine.*

17. Board of Health quarantine notice, San Francisco, circa 1910. *Courtesy of National Library of Medicine.*

18. AIDS virus. *Courtesy of Centers for Disease Control and Prevention.*

19. AIDS virus particles found in a patient with hemophilia. *Courtesy of Centers for Disease Control and Prevention.*

20. Gerald Ford's nemesis, the swine flu virus. *Courtesy of Centers for Disease Control and Prevention.*

21. Human liver infected with the Ebola virus. *Courtesy of Centers for Disease Control and Prevention.*

22. A Centers for Disease Control worker exposed to the Ebola virus. *Courtesy of National Library of Medicine.*

23. A Japanese woodcut, circa 1900, showing measles as an especially hirsute demon. *Courtesy of National Library of Medicine.*

24. Rubella virus. *Courtesy of Centers for Disease Control and Prevention.*

25. Mumps virus. *Courtesy of Centers for Disease Control and Prevention.*

26. "A Memory of the Mumps," *Harper's Weekly,* 1860. *Courtesy of National Library of Medicine.*

27. "Penicillin Cures Gonorrhea in 4 hours." *Courtesy of National Library of Medicine.*

28. "Gonorrhea, the Great Sterilizer." *Courtesy of National Library of Medicine.*

29. Malaria poster. *Courtesy of National Library of Medicine.*

30. Red blood cells infected with plasmodium. *Courtesy of Dr. Edward J. Bottone.*

31. "The War with Malaria," a Soviet educational poster. *Courtesy of National Library of Medicine.*

32. Board of Health quarantine notice, San Francisco, circa 1910. *Courtesy of National Library of Medicine.*

33. Elvis gets his polio shot. *Courtesy of March of Dimes Birth Defects Foundation.*

34. A cartoon published in 1826 shows a rabid dog causing panic in the streets of London. *Courtesy of National Library of Medicine.*

35. Sopona, god of smallpox, worshiped by the Yorubas of Nigeria. *Courtesy of World Health Organization.*

36. Shitala Mata, Indian goddess of smallpox. *Courtesy of World Health Organization.*

37. Lady Mary Wortley Montagu (1689–1762). *Courtesy of National Library of Medicine.*

38. A milkmaid's hand. *Courtesy of National Library of Medicine.*

39. An antivaccination society illustration, London, 1802. *Courtesy of National Library of Medicine.*

40. *Staphylococcus* bacteria. *Courtesy of Dr. Edward J. Bottone.*

41. Twisting chains of *Streptococcus* bacteria. *Courtesy of Dr. Edward J. Bottone.*

42. Pneumococcus cells. *Courtesy of Dr. Edward J. Bottone.*

43. "Juke Joint Sniper." *Courtesy of National Library of Medicine.*

44. French TB poster. *Courtesy of National Library of Medicine.*

45. "Your Lungs," Germine Company TB poster. *Courtesy of National Library of Medicine.*

46. "The Next to Go," Red Cross Christmas Seal Campaign TB poster. *Courtesy of National Library of Medicine.*

47. Chicken pox virus. *Courtesy of Centers for Disease Control and Prevention.*

48. *Ascaris lumbricoides,* or roundworm, infects a billion people worldwide. *Courtesy of Dr. Edward J. Bottone.*

49. An *Aedes aegypti* mosquito, first identified as the carrier of yellow fever in 1881. *Courtesy of American Museum of Natural History.*

50. A nineteenth-century cartoon showing Columbia rescuing Florida from Yellow Jack. *Courtesy of Library of Congress.*

51. Plague bacteria. *Courtesy of Dr. Edward J. Bottone.*

52. Scientists examining rats for bubonic plague in New Orleans, 1914. *Courtesy of National Library of Medicine.*

53. "The Plague Pit," an early-nineteenth-century engraving. *Courtesy of National Library of Medicine.*

54. An angel tells Charlemagne of a plant that can stop the Plague. *Courtesy of National Library of Medicine.*

A
Field
Guide
to Germs

An Egyptian stele, or stone tablet, from the eighteenth dynasty (1580–1350 B.C.), depicting a man with a withered leg characteristic of polio.

· *Introduction* ·

After a decade that began with Americans afraid to open their possibly infectious mail and ended with dispensers of hand sanitizer as common around college campuses as cans of beer, germs continue to fuel our daily worries. The anthrax-laced letters of 2001's bioterrorism scare proved to be a scurrilous episode of limited scope, but the swine flu epidemic of 2009 put millions of people in bed and killed significant numbers around the world. The first story dramatized how some microbes can be "weaponized" to harm people by a criminal or simply mad intelligence—not a new development in history, but still a terrifying one. The second reminded everyone that nature's kitchen never stops serving up fresh organisms that can make us sick if we encounter them by chance.

Rich nations devote vast portions of their economies to medical practice and associated work. About one-sixth of *all* the money spent on goods and services in the United States goes to health care: more than two trillion dollars, which is a figure difficult to grasp. Fifty years ago the fraction was one-twentieth, which means that while the whole pie has grown since then, medicine is taking a much bigger bite of it. (It is ironic that perhaps the only other industrial complex besides medicine that has garnered such bedrock support is the military, which also maintains a cloistered profession claiming skills that a fearful society should not be without.) Meanwhile, the world's poor must survive without the fruits of this investment, preoccupied as they are with mundane problems like no sanitation, no pharmaceuticals, and no doctors at all.

For anyone able to buy and read this book, wickedly infectious microbes will continue to be rather unusual. The goal here is to sort out what is worth worrying about from what is merely fodder for those who profit from anxiety. We always have and always will live in a sea of countless invisible organisms—most of them harmless or even beneficial—that outnumber us by many orders of magnitude.[1] We depend on them, actually. Those that might hurt us are as vulnerable as ever to medical science and, in varying degrees, our own immune systems. Indeed, there is a broad shift taking place in "morbidity and mortality," as the professionals call sickness and death, from infectious to noncommunicable diseases like cancer.[2] A field guide is thus offered as a bulwark against phobia and a reminder that medicine is a socioeconomic activity.

It is useful to review some basics. The pejorative "germ," from a Latin word meaning "sprout," is not a precise scientific term. The microscopic (and submicroscopic) creatures of concern are mostly bacteria, living single-cell parasites that may cause disease through runaway infection of their host or the release of powerful poisons, and viruses, which are inert by themselves but have the nasty ability to pirate a host's metabolism. (Two categories of more complex organisms, fungi and protozoa, are of somewhat narrower interest.) The ones that can hurt us are called pathogens, from the Greek word for disease, *pathos*. Antibiotics slow the growth of pathogenic bacteria without zapping our own cells.[3] Since there are precious few antiviral drugs that will not also destroy the infected host, we depend on vaccines—usually made from safe pieces of the bug that nonetheless trigger our defense network—to combat viruses.

It took thousands of years of very roughshod medical practice and seat-of-the-pants science to arrive at this rubric. Dutch scientist Antonie van Leeuwenhoek (pronounced "layvenhook"), a Delft draper who became fascinated with lens making, reported on his observations of microorganisms in fresh water beginning in 1674.[4] He called them "animalcules." Though a Roman writer named Varro had speculated during the first century B.C. about whether people got sick after inhaling "tiny animals" (one can imagine his friends imploring him to lay off the grape), belief that such parasites came from nowhere—that is, that they were generated spontaneously inside the body—prevailed well into the nineteenth century. A true germ theory of disease was

French chemist and microbiologist Louis Pasteur (1822–1895), known for gentleness toward small creatures, with his granddaughter.

spearheaded by the study of fermentation by the father of modern bacteriology, French chemist Louis Pasteur (1822–1895)—he was interested in the wine-making process, of course—beginning in 1857 and shown in a paper on wound infections published by the pioneer German bacteriologist and Nobel Prize winner Robert Koch (1843–1910) in 1879. In general, the theory held that a pathogenic organism must be present in every case of the disease; that this organism could be grown or "cultured" outside the infected body; that an animal inoculated with this culture would get the same disease; and that the organism could be found again in the inoculated animal and recultured. Some pathogens may not meet all of these conditions, but they are the boilerplate for explaining scientifically how disease happens.[5]

For eons, we and the bugs have lived in a state of largely peaceful coexistence, modulated like everything else by natural selection. Of course, a great deal of selection has already occurred—we are here because our ancestors somehow avoided deadly onslaughts. But different individuals and whole communities still show various degrees of resistance to infection by the relatively few organisms that are pathogenic. This can sometimes feed a scapegoat mentality for attaching blame, as when syphilis was called "the French disease" in Spain and "the Spanish disease" in France.[6] It also explains why even the deadliest germs are not 100 percent lethal.

Our defenses, which are formidable, are constantly adjusting. And so are the microbes, whose mutability is astonishing. The bugs have one big advantage: New generations may take only a few minutes to mature, compared to decades for our offspring (if we're lucky). Such interaction, generation after countless generation, creates a web of mutual adaptation that permits both host and parasite to survive. The most perfectly adapted parasites cause little or no damage, thus escaping the notice of everyone except scholars. If a disease organism kills lots of people very fast, however, it causes a crisis all around, because

German bacteriologist Robert Koch (1843–1910) and frau in Japan, 1903,
where he was revered as a god.

new hosts must be found often enough to keep its progeny
going. In this sense, many of the most virulent germs can be
thought of as gluttonous savages that are still in the early stages
of biological adaptation (they may never be benign, just less dis-
ruptive). The time scale involved for a new pathogen to appear
on our horizon is considerable: What we perceive as new or

"emerging" pathogens are generally not straight out of nature's kitchen, but existing microbes that found us through changes in the environment or human ecology.[7] As Luc Montagnier, the Pasteur Institute virologist who isolated the AIDS virus, once said, "We're boarding a train that's already in motion."[8]

Infections, like civilizations, rise and fall. Small comfort, since the balance of power is constantly changing. For example, after penicillin became widely available in the 1940s, most bacterial infections were treatable. By the 1960s, doctors were getting cocky about totally vanquishing infectious disease. But the evolutionary clock never stops ticking. Now, strains of common diseases have developed that are resistant to penicillin. Given enough time, new microbial strains are guaranteed to appear naturally on which many current antibiotics have insufficient effect. This has already been noticed with certain relatively rare bacteria, but they may share their genes for resistance with other species that are more common. Most pathogenic bacteria are by now resistant to at least one antibiotic, a few to more than one. Once a resistant strain appears in a few scattered places, it can spread around the planet in a matter of months. Overuse of antibiotics, in livestock as well as humans, speeds up the process by wiping out the unresistant competition.

Many researchers fear that the antibiotic spree in the decades after World War II was a blunder, creating microbial freaks that assume unnatural dominance. The same dynamic applies to viruses under prolonged treatment. Recent studies have found that as many as 16 percent of people newly infected with human immunodeficiency virus (HIV) in the United States harbor a strain resistant to at least one of the antiretroviral drugs used to combat AIDS. The probability of developing resistance after six years is more than 25 percent. Thus, in the wake of the widespread use of combination drug therapy that revolutionized AIDS treatment in the 1990s and saved tens of thousands of

Fresh AIDS virus budding from the membranes of infected T-4 lymphocytes.

lives, many people now face the prospect that their infections will become increasingly difficult to control.

There are, to be sure, diseases that fly from victim to victim with the greatest of ease. The once common afflictions of childhood (common, at least, without vaccinations)—measles, chicken pox, sometimes deadly smallpox (now extinct in nature), whooping cough, mumps—have at times been nearly as familiar as the sniffles. Happily, a single bout with them usually produces lifelong immunity. But if some of these same viruses were to invade adult populations with no previous exposure, we could see the same level of destruction that helped a handful of Conquistadors to topple the Aztec and Inca empires, or that decimated North American Indian tribes by as much as 50 to 90 percent after contact with Europeans. Hence the concern about smallpox as a bioterrorist weapon, no matter how unlikely the scenario.

Today, at least in developed countries, the most serious threat of infection comes not from virulent germs that exist outside our bodies, but from the myriad parasitic organisms we play normal host to (known rather daintily as our "resident flora"). They may live on our skin, in our mouth, in our throat or lungs, in our stomach or intestinal tract, and in our genitalia.[9] They are we. Since they can make us sick, depending on the condition of our defenses, we are constantly warding off infection as a way of life. In a sense, we are all exposed, all the time.

When the poet Baudelaire gazed upon the Paris barricades in 1848 and proclaimed, "We are democratized and syphilized," his sentiment, albeit antirepublican, was an accurate estimation of the new order. "The rationality of life is identical to the rationality of that which threatens life," the French philosopher and historian of science Michel Foucault insisted 115 years later.[10] It is a rare individual who can embrace this tenet under duress, but the principle makes a good armature for personal attitude and public policy.

In the United States, medical duress goes hand-in-hand with financial extremus for much of the population. Some 45.7 million people lacked health insurance coverage in 2007, the most recent year for which Census Bureau data is available, with 2.3 million more predicted to lose it every succeeding year. The majority of uninsured are from working families who cannot obtain insurance either because their employer does not offer it or because they cannot afford what is available. Nearly one out of four non-elderly Americans *with* insurance—53.2 million people—spent more than 10 percent of their pre-tax income on health care in 2009, and 14.3 million spent more than 25 percent. The American medical system is the most expensive in the world, yet it consistently underperforms relative to other countries in most measurable dimensions.[11] In one of the richest

An illustration from a poetry book, published in Paris by Auguste Marseille Barthélemy in 1851, showing a young suitor kneeling before Death disguised as a beautiful girl. That syphilis was frequently passed in the opposite direction would not have been fit to portray.

nations, just as in the poorest, getting sick can mean impoverishment.

Needless to say, medical practice, like auto repair or the merger and acquisitions bar, is the bailiwick of high-pay specialists. This seems to be broadly accepted by well-off Americans (while in Europe, where national health insurance prevails, even the British Medical Association opposes up-front payments to doctors on the grounds that this would damage the doctor-patient relationship and hit the poor hardest). The way it looks is that the buyers of medical care lost pull to the sellers quite some time ago. Big-time capitalism may be partly to blame, with its penchant for giving consumers lots of choices but rarely any power other than the ability to reject inferior merchandise. Acquiring the services of a doctor is a singular market transaction, where the weakness of one party contrasts utterly with the stature of the other. "There are only two classes of mankind in the world— doctors and patients," Rudyard Kipling quipped. Just as myriad scientific advances have widened the gulf between developed and Third World nations, access to medical expertise has grown more restricted for the poor.

Yet medicine has always been a cloistered subculture where unique responsibilities foster aloofness.[12] Add to this the fact that no profession ever had a better price maker than the Grim Reaper and there is obviously an ineluctable potential for abuse, the Hippocratic oath notwithstanding. During the Middle Ages, costumed not in white lab coats but purple or red gowns with furred hoods, physicians (and their wives) were permitted extra luxury by sumptuary laws. Educated professionals were for the well-to-do. Jean Froissart, the great fourteenth-century historian, noted in his *Chronicles* that "the aim which doctors always pursue" was "to get large payments from the lords and ladies they attend." To distinguish themselves from motley barbers, pig gelders, bath attendants, and other semi-charlatans who actually dealt face-to-face with masses of sick people, university-trained

A woodcut from Thomas Bartholin's Historiarium anatomicarum *(1654–1661) showing a doctor garbed against the Plague. The beaked mask, perhaps the original nosegay, contained pleasant-smelling stuff to cover the reek of death.*

physicians in France swore an oath that they would never perform surgery. When confronting apocalyptic ills like the Black Death, which ultimately killed a third of Europe's population, concoctions of powdered emeralds and gold dust played straight to the wealthiest patient's timeless sense that therapeutic value varies directly with cost.[13] ("If you want to be cured of I don't know what, take this herb of I don't know what name, apply it I don't know where, and you will be cured I don't know when," went one medieval piece of jocosity. Leonardo da Vinci, no lowbrow, warned others to "keep clear of the physicians, for their drugs are a kind of alchemy concerning which there are no fewer books than there are medicines.") Judaism taught that the physician was an instrument of Providence.[14] Universal prescription of opium into the first decades of the twentieth century for everything from the blues to fever must have helped make medical men seem angelic indeed. Only in the last several generations has the profession's practical foundation become very much deeper than the feel-good psychology of letting a self-confident *Herr Doktor* take control in vital emergencies.[15]

Like the folkways they evolved from, medical theories can still be dogmatic. Some, like the American Psychiatric Association's view until 1973 that homosexuality was a disease, are derived from rank prejudice. Others, like the overabundance of cesarean sections and coronary bypass operations, reflect nonscientific influences. Few establishments are capable of mustering more disdain against criticism of their orthodoxies, as though the complainant were obstructing Western triumphalism itself. Because the practice of medicine is not a science, but the application of many branches of science to the way people live (thinking of doctors as skilled craftsmen can be revelatory), it harbors far more subjectivity than state licensing boards might want to admit. Hence the common advice to "get more than one opinion," essentially to shop around. What remains the same for shoppers, of course, is the crushing fear of the unknown, the humiliation and depersonalization of having no clue whatsoever about nature's exigencies, the insult of seldom finding anyone glad to lay out a few fundamentals.

Despite decades of academic debate about a new ethics to bridge the social gap between doctor and patient, there is still a sizable reactionary force behind the notion that the power to heal resides at least partly in the gap itself. (We should keep in mind that the term *informed consent* was coined only in 1957, and then as a legal weapon.) According to this venerable way of thinking, most people are barely capable of assuming freedom and responsibility in medical matters when they are well. When they are sick, they need the "miracle, mystery, and authority" of a paternalistic physician.[16] The so-called placebo effect—getting better after taking a fake pill—is thought to be evidence of this desire, but even bald-faced lying by a physician finds a time-honored place here, where "bad news" is felt to produce further loss of health.[17] A doctor's personal charisma and social clout are thus just as important, and sometimes more important, than medical knowledge, which is anyway imperfect.[18] "The presence of the doctor is the first part of the cure," says a French proverb. Better that the hopeful supplicant remain ignorant of medicine's limitations, or so the rather patronizing, self-serving argument goes.

At the same time, there has developed an unsavory congruence between this condition of ignorance and the indecipherable excess of doctor bills. Staff parking lots at major hospitals filled with luxury cars do not send a reassuring message. As the perception of rip-off has risen, so has a general uneasiness about catching something. Economics is certainly a factor here, but probably a minor one compared to the slide of baby boomers into middle age, the decay of big cities into Third World poverty, and the existence of a worldwide plague that continues to defy the power of research. Simultaneously, then, there are intimations of mortality among a vast cohort that never quite believed it would grow old, social policy failures in the very places where everyone was supposed to find the best opportunities, and an epoch of death-by-sex on the heels of the most bacchic sexual revolution in Western history. Small wonder that a certain edginess about contagion has crept into daily encounters of all kinds.

It could be a lethal mistake, of course, to assume that what constitutes "good practice" should pivot upon a patient's freedom of choice (what the ethicists refer to as "patient autonomy"). Medical diagnosis is not for hobbyists, any more than launching a space shuttle is for astronomy buffs. Doctors will continue to hold a tight monopoly on technical skills, even as their decisions are scrutinized through managed-care systems.[19] A worshipful society, moreover, is unlikely to adjust their socioeconomic status earthward. But signing a blank check is seldom a wise idea and never a healthy feeling, even for hypochondriacs. There will always be the lucky well and the unlucky sick, but the division should never be determined by privilege—intellectual or otherwise.

Among the myriad tragedies of the AIDS epidemic has been the way it played into the hands of a puritan element who would seek to proscribe human nature under any circumstances.[20] The horrific list of previously arcane diseases associated with the syndrome that have now become part of the lingua franca has helped fuel a Social Darwinist attitude toward homelessness, poverty, or unconventionality in general. This may be a primordial reflex—the Latin root of the word "infected," *infectio,* means "tainted, spoiled, or polluted." Similar clashes occurred during the American cholera epidemics of the nineteenth century, when only disreputable members of society seemed to catch the disease. The imagined contagiousness of their vice—be it poverty, Irishness, or some other moral failing—was as feared as their sickness.[21] Immigrants have traditionally been subject to such medicalized nativism.[22] Close analogies can also be found during the plague years of ancient and medieval history, when the stress of mass death altered the integrity of church and state. Beginning in 1348, thousands of European Jews were murdered for supposedly poisoning wells with Black Death. In 1630, at the start of an epidemic that

"Fear of disease causes the disease of fear." An early-nineteenth-century caricature
of hypochondria and pseudoprotective measures against cholera.

would wipe out more than 30 percent of Venice's populace,
laws were enacted to clear beggars from sacred quarters of the
city. This was intended as an act of public atonement, not
hygiene.[23]

The triumph of scientific medicine in the last century was to
push calamitous encounters with pathogenic organisms away
from bourgeois life. Deadly or disabling illness of this sort is
more closely associated than ever with poverty, malnutrition,
unconventional lifestyles (this is a *very* loaded phrase, but it
will have to do, the professional alternative being "special socio-
behavioral features"), sexual promiscuity (no moral judgment

Trying to make the best of quarantine on Ellis Island, New York.

intended, just arithmetic), substance abuse, and other facts of human nature that tend to be ignored by polite circles whenever possible. Poverty has always put people in harm's way. Many of history's great epidemics have shown strong "socioeconomic factors," as modern epidemiologists delicately say. The fact that a new flu virus seems to negate progress, throwing everyone back into the not-so-distant past when the well-to-do were no better prepared to fight off a germ than the destitute, may account for some of its psychological force.

The bourgeoisie should never grow smug, of course. As existing infections are pushed to the margins, new ones will fill the space as evolution's imperative asserts itself. Flu viruses can be stymied but nobody can escape the ecosystem.

Nor can anyone escape the need for a well-rounded edu-

In an 1883 cartoon, Science sleeps on a New York dock as Cholera floats across the ocean from Europe.

cation. There is room for humor and seriousness, history and science, government and philosophy. Here, placed in a common context, are the top-ranked terms and germs (in prevalence, or power, or worry factor, or even literary interest) of the billions of potentially unfriendly organisms out there, presented not with the quack promise of self-diagnosis, but with the absolute certainty that a little knowledge is always better than zip. Put away the disinfectant soaps and sprays, turn off the television news, and study the ones that infect the imagination more easily than the body. All of them, in other words.

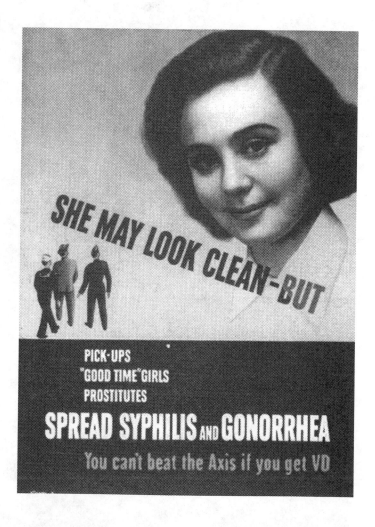

• adenoviruses •

The adenoviruses are a family of fifty-one viruses, identified by sequential letters and numbers, first found in adenoids (lymphoid tissue at the back of the pharynx) in the early 1950s, but definitely around for a long time before that. There are also adenoviruses that infect only animals. The ones that like people cause about 5 percent of all respiratory illnesses, from mild flu-like symptoms to pneumonia, which are rarely fatal though of special concern among military recruits. Some types cause gastrointestinal illnesses; others can cause conjunctivitis among swimmers in lakes or insufficiently chlorinated swimming pools. Most children around the world have been infected with the more common adenoviruses by the time they reach school age.

Outbreaks of nontrivial respiratory illnesses in new soldiers have been a recognized problem since at least the Civil War. The term "acute respiratory disease" gained acronym status as ARD during World War II. Up to twenty cases a week per one hundred recruits could be expected, creating a significant drag on training. Starting in 1971, all American military recruits were vaccinated against the most common adenoviruses, but in 1996 the sole manufacturer of the vaccine stopped production. By

1999, illness during basic training had returned to pre-vaccine-era levels, and in 2000 the first two deaths occurred since 1972.

In 2006 and 2007 a rarely reported adenovirus type called Ad14 caused a cluster of 140 cases of severe respiratory illness, with nine fatalities in Oregon, Washington, and Texas (at Lackland Air Force Base), as well as one infant death in New York City. This suggested the emergence of a new and virulent brand of Ad14. Despite the constant attention of a National Surveillance for Emerging Adenovirus Infections system that includes military and civilian labs, the control of outbreaks can be difficult to effect because the viruses are carried in both respiratory secretions and feces and because they can live for weeks on environmental surfaces.

Just why some of the adenoviruses enjoy these young soldiers so much, compared to civilians, is not clear, but crowding and stress are clearly part of the problem. Populations of recruits have traditionally been subject to diseases usually seen in childhood, such as chicken pox, mumps, and rubella (German measles), perhaps because of the mixture of immune and nonimmune bodies from many different locales. Except during massive mobilizations, troops can be separated into small units when they first arrive, which helps to avert epidemics. Then again, many of these guys and gals don't mind a few days in sick bay.

The ability of adenoviruses to infect our tonsils and adenoids for long periods of time implies that they have learned how to outwit our defenses. They do not fall into a latent state, like herpes viruses, but reproduce constantly at a snail's pace, changing the cells they enter in such a way that the immune system cannot tell the difference. Some of their tricks give them the power to produce tumors, but this has been observed only in experimental hosts such as hamsters. Why adenoviruses cannot cause cancer in people is unknown.

• anthrax •
(Bacillus anthracis)

When the Lord visited "a very severe plague" upon Pharaoh's cattle in Exodus 9, it was surely anthrax. Tough *Bacillus anthracis* spores can persist for decades in alluvial soil like that of the Nile valley, ravaging herds no matter whose side God is on—in this case, the Israelites were probably spared because they camped on sandy ground above the river's floodplain. Livestock dying in the grip of anthrax's gruesome spasms and convulsions would definitely seem cursed. (The lower Mississippi River valley is another well-documented haven for the disease—the first American cases were reported among animals in Louisiana in the early 1700s.) Anthrax is known as a *zoonotic* disease (from the Greek prefix *zo-,* meaning "life" or "animal," and *nosos,* "disease") because people catch it from animals—primarily cattle but sometimes sheep, horses, pigs, or goats—not from other people. Spores enter the body by means of cuts (cutaneous

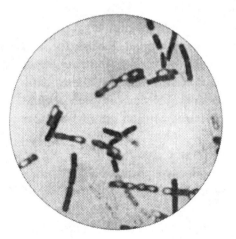

Anthrax cells, the first bacteria ever seen, magnified a thousand times.

anthrax, which causes the coal-colored skin lesions for which the microbe is named, from the Greek *anthrakis,* meaning "coal"); the breath (inhalation or pulmonary anthrax, rare but quite deadly); or ingesting bad meat (gastrointestinal anthrax). The spores then multiply and produce enough toxin within a week or so to injure surrounding cells and tissue. Today, of course, we know anthrax as bioterrorist fan mail.

In mid-nineteenth-century England and Germany, anthrax was called woolsorter's and ragpicker's disease, respectively, because these workers caught it from spores in fibers and hides. Robert Koch helped launch the science of bacteriology in 1876 by developing a method to grow pure anthrax culture in the laboratory. When anthrax spread through sheep herds in France in 1877, Louis Pasteur began research that produced the first successful vaccination of livestock in 1881. Laws were enacted in the 1920s to require testing of shaving brushes that used horsehair or pig bristles. The largest outbreak in the United States happened in 1957, when nine employees of a goat hair processing plant became ill after handling a contaminated shipment from Pakistan. Four of five patients with the inhalation form of the disease died. In the 1970s, cases were traced to contact with souvenir drums from Haiti fitted with goatskin drumheads. An amateur weaver died after breathing spores in imported yarn. Doctors recorded hundreds of cases of cutaneous anthrax in the United States in the twentieth century, chiefly among agricultural workers; only eighteen cases of inhalation anthrax were noted during that time, with the last reported in 1978.

Because the spores are so efficient, anthrax has long been researched for use as a weapon. In 1925, drawing on the language of World War I peace treaties, the Geneva Protocol banned bacteriologic warfare and was ratified before World War II by all the great powers except the United States and Japan. After the war, President Truman withdrew the treaty from the Senate. A Soviet resolution calling on all United Nations members to ratify the ban was rejected in 1952. From then on, the United States and the Soviet Union, as well as other major nations, maintained extensive facilities for producing and testing germ weapons, including anthrax. In 1969, following an incident in which nerve gas drifted off a secret military test range in Utah and killed hundreds of sheep, President Nixon renounced chemical and biological weapons. In 1975, the United States finally ratified the international ban on such armaments.

A 1979 anthrax epidemic in Sverdlovsk (now Yekaterinburg) in the Soviet Union raised concern about whether the outbreak was natural or due to activities at a military microbiology lab. Work on offensive biological weapons was illegal, but defensive development of vaccines and protective gear was permitted by the treaty. Soviet officials claimed at the time that victims developed gastrointestinal anthrax after eating contaminated meat and cutaneous anthrax from contact with infected livestock. In 1992, Russian President Boris Yeltsin announced that the KGB had admitted that the outbreak had a military cause. Two years later, an independent team of American and Russian experts determined that a windborne aerosol of anthrax spores that had escaped from a military lab on April 2, 1979, produced the epidemic that killed sixty-eight of the seventy-nine cases, making it the largest documented outbreak ever of human inhalation anthrax.

Bioterrorist-related cases of cutaneous and inhalation anthrax in the United States in 2001 (including five deaths from inhalation) have been widely reported in the popular and professional media. Readers of all levels can stay abreast at the Centers for Disease Control and Prevention's Web site (www.cdc.gov), thereby avoiding the hysteria that understandably accrues to this subject and is, indeed, part of the psychological intent of terrorism. The book is being rewritten, so to speak, on the pathogenesis, diagnosis, clinical manifestations, therapy, epidemiology, and decontamination of anthrax. Even the number of inhaled spores necessary to produce disease—previously based on animal tests at military labs—is being reconsidered in light of what happened to postal workers and others. Most of the military scientific information from decades past, especially regarding so-called weaponized germs like the material released in U.S. Senate offices, was secret and therefore unavailable or is of dubious value to current researchers and physicians.

For the general public, the risk of getting anthrax remains negligible, at least in developed countries with modern animal husbandry and industrial hygiene. The greatest fear is fear itself, now as ever. An anthrax vaccine containing no dead or live bacteria that is about 93 percent effective has been licensed for humans. It is recommended only for some military personnel and certain workers in close contact with animals or animal products.

• arboviruses •

The word "arbovirus" derives from the phrase "*arthropod-borne virus*," which means viruses that propagate inside insects and other arthropods and reach us through bites. There are more than 520 known arboviruses, of which about a hundred cause disease in humans, usually with no apparent symptoms. But encephalitis, yellow fever, dengue fever, and a veritable zoo of exotic tropical fevers and malaises—Mayaro, Kyasanur, Bunyamwera, Marituba, Punta Toro, Candiru, West Nile—give these microbes a deservedly bad reputation. Mosquitoes, ticks, and sandflies are the most common carriers, with various other flies and mites guilty to a lesser degree.

People are usually "dead-end" hosts for the arboviruses. That is, most of the germs don't need us for long-term survival—we just get in the way sometimes and then feel sick. Birds are much more crucial to arboviruses as long-term hosts. The exceptions of great significance are yellow fever, dengue, and chikungunya, where we serve as a vital link in their life cycle. Before the jet age, many of the arbovirus illnesses were encountered only by Indiana Jones types, but today they can turn up almost anywhere, at least in isolated cases. Epidemics are unlikely wherever the associated insects are kept under control.

• arenaviruses •

Arenaviruses—named after the Latin word for sand because of grainy ribosomes (protein factories) they capture from host cells—are a family of viruses first identified in 1933 during an encephalitis outbreak in St. Louis, Missouri. They turned out to be innocent in this epidemic, but one was found to cause meningitis. Members of the tribe that have so far been identified sport names that would make a powerful magical incantation: Amapari, Junin, Lassa, Latino, Machupo, Guanarito, Sabia, Parana, Pichinde, Tacaribe, Tamiami. Their natural hosts would spike the brew even more: bats, rats, and mice. Some of these viruses bring on meningitis or various hemorrhagic fevers when humans come into contact with infected excreta. *Feh!*

They can be quite deadly. Argentinian hemorrhagic fever, caused by the Junin virus discovered in 1953 near the Junin River, may produce a 20 percent mortality rate as it disrupts capillaries and essentially causes its victims to bleed to death. The Bolivian brand (Machupo, also named after a river), first called *el typho negro,* the black typhus, can reach 30 percent mortality. It probably emerged via field mice of the genus *Calomys* as jungles were cleared for agriculture during the 1950s. Lassa fever, which first came to the attention of Western scientists when American nurses were stricken in a Nigerian village of that name in 1969, topped 60 percent mortality in one hospital outbreak. Patients are therefore attended by medical staff in biohazard moon suits— not a reassuring sight at bedside. Disinfection procedures under such circumstances are called "barrier nursing," a term that neatly captures the scene.

Fortunately, most of these killers have remained confined to certain geographic areas, though cases have appeared elsewhere through the swift vagaries of modern travel. In 1999 and 2000, three fatal cases of indigenous arenavirus infection occurred in California involving Whitewater Arroyo, a strain first detected in New Mexican woodrats in the early 1990s. One victim apparently cleaned rodent droppings at home just before getting sick, but the other two had no such history of exposure.

Direct person-to-person contagion is the exception, not the rule, so with proper surveillance these diseases should stay in the exotic category. Unfortunately, they most often occur in parts of the world where effective monitoring may be too much to expect of state medical agencies.

• **bites** •

More than 4.7 million Americans are bitten by dogs every year, accounting for hundreds of thousands of emergency room visits and many millions of dollars in direct medical-care costs. Most nonrabid bites are provoked, and most victims are males under twenty years old, with the highest rate occurring in children between ages five and nine. Nearly a third of these wounds may become infected, regardless of first-aid measures like washing or applying salves. In general, bites to the hand are more serious and more frequently become infected than chomps to other areas. Infections are due mostly to germs in the dog's mouth, including *Pasteurella,* staphylococci, and *Weeksiella.* Snakes account for another eight thousand punctured Americans annually. But human bites, including those self-inflicted by finger suckers, are far worse than animal bites. Many of the bacteria in human bite wounds may be resistant to penicillin. The bacteriology of Clenched-Fist Injuries, where one bloke socks another in the teeth, is similar to that of bites and should be taken into consideration by Hollywood screenwriters. Vampires are already big money in Tinseltown, of course.

• *Bordetella pertussis* •
(whooping cough)

One of the planet's great ubiquitous organisms, the bacterium that causes whooping cough, *Bordetella pertussis,* cannot be avoided. It is everywhere and ferociously virulent, infecting 90 percent or more of susceptible individuals after exposure through close respiratory contact. It thus strikes early in life, with fatality rates highest among the youngest—a hundred years ago it was a major cause of infant mortality around the world, and it may still grimly reap wherever people are too poor or isolated to get immunized. It causes about 40 million cases of whooping cough (pertussis) and 400,000 deaths annually. A vaccine was widely available in the United States by the mid-1940s. Pertussis deaths began to decline for unknown reasons even before that decade, but local outbreaks still occur and have been increasing in the United States since the 1980s, perhaps due to lax immunization. In 2007, 10,454 cases were reported and ten children died. Epidemics arise every three to five years—in 2005, reported cases spiked at 25,616. In 1999, there were 7,288 cases reported in the United States, 11 percent of which were among kids between one and five years old whose immunity should have been assured. Adolescents and adults are susceptible as immunity from childhood vaccination weakens.

The terms "pertussis" and "whooping cough" first appeared in English in the seventeenth century. Before that the disease had been called "chin cough" or, in France, *quinta* (apparently referring to an interval of about five hours between coughing spasms) and *coqueluche* (also used for influenza). After incubating for one to two weeks, *Bordetella pertussis* invades the respiratory tract, causing a hacking cough. Over the next ten to fourteen days this progresses into a paroxysmal cough that may happen five to fifteen quick times in a row, followed by a high-pitched, crowing deep breath and vomiting. Convalescence begins within a month, and most people recover after a seven-week illness. Pneumonia and brain damage are possible complications. Whooping cough was epidemic in the American colonies, particularly in South Carolina. *Bordetella pertussis* was not isolated until 1906, by the Belgian physiologist and Nobel Prize winner Jules Bordet (1870–1961).

Fighting the germ requires diligent effort. Beginning when they are two months old and continuing at intervals until they are seven years old, children in developed countries usually receive five doses of vaccine (a combination of diphtheria, tetanus, and pertussis vaccines, or the DTaP shot—the "a" stands for "acellular," which means it contains only a fragment of the *Bordetella pertussis* cell). About 85 percent of American children receive this regimen. Another combination called Tdap is given as a booster to adolescents and adults. The old DTP shot that contained whole dead bacteria and was used starting in 1940 is no longer given in the United States.

• *Borrelia burgdorferi* •
(Lyme disease)

Lyme disease, caused by the spirally coiled bacterium *Borrelia burgdorferi,* is the most common insect-bite ailment in the United States and Europe. About 20,000 cases a year are reported in the United States, where it occurs almost everywhere but especially in New York, Connecticut, Pennsylvania, New Jersey, Rhode

Island, Massachusetts, Delaware, Maryland, and Wisconsin. Though this malady has caused unnecessary hysteria in some areas (enough to give it the nickname "borreliosis neurosis"), it provides a textbook example of how epidemics may seem to pop out of nowhere as a result of subtle changes in the environment and human ecology.

By the early 1800s, vast tracts of land in the eastern United States had been denuded for firewood, lumber products, and farming. As virgin forest was replaced with an utterly different habitat of fields or low brush, the larger mammals that had thrived for eons all but disappeared, notably deer and their predators. By the mid-1800s, however, the process started to reverse as agriculture moved westward. Abandoned farmland did not turn back into the relatively open growth of great trees that characterizes wilderness, though, but into scrubby woodlands choked with undergrowth. By 1980, there was four times more wooded acreage in the Northeast than there had been in 1860, but in biological terms it was a brand-new environment, a wide-open ecological niche. Deer began to return in the early 1900s (mostly without their old predators, which were hunted to near extinction) and reached record levels by the end of the century. Also at record levels, of course, were the numbers of humans now living in close proximity to the deer. The conditions were perfect for a vicious circle of infection that shows no sign of being broken.

The first recognized outbreak occurred during 1975–1976 around the town of Old Lyme, Connecticut, where clusters of arthritis cases among children attracted medical attention. Retrospective studies have identified a few apparent cases in the 1960s in Wisconsin and Massachusetts and as early as the 1920s in forested parts of northern Europe. *Borrelia burgdorferi,* named after Willy Burgdorfer, the scientist who identified the species in 1981, reside harmlessly in white-footed mice (*Peromyscus leucopus*), which just happen to be the preferred feeding site for larvae and nymphs of the deer tick, *Ixodes scapularis*—formerly called *I. dammini,* a suitable name, since they are about the size of poppy seeds and damnably hard to see. (In California, *I. pacificus* ticks are the villains, and their menu involves lizards.) After the baby ticks eat mouse blood for about two days in the summer and four days the following spring, they become infected with *B. burgdorferi.* At the nymph stage they can infect humans. Adult ticks would rather suck deer blood, however, and it is at this stage that they may hop off a deer and soon chance upon humans who brush against infested vegetation. This cycle can be very efficient. *I. scapularis* ticks were first discovered on deer carcasses in Ipswich, Massachusetts, in 1979. By 1986, the annual incidence of Lyme disease there exceeded 10 percent of the population, though this figure was probably inflated by overdiagnosis. There is no evidence for person-to-person contagion.

What can be done? Besides controlling deer populations—which is often a controversial approach and one that may not decrease tick abundance—or wearing tick repellent and long clothes in the spring and summer months in addition to being sure to shower after a day outdoors and to check for attached ticks (transmission of the disease does not happen unless the tick stays on for thirty-six to forty-eight hours) . . . not much. In about three-quarters of people who catch Lyme, a blotchy red rash, often circular, will surround the bite within thirty days, and flu-like symptoms will occur. But others may not notice anything abnormal for months while more serious conditions develop. In Lyme-disease hot zones, a preventive dose of antibiotic after the discovery of an attached tick might be a good idea. No vaccine is currently available.

In the northeastern United States, between 15 and 30 percent of all deer ticks are infected (or about half of all adult deer ticks). In most areas, only 1 to 3 percent of people who get bitten actually catch the disease, which is usually easy to treat with antibiotics if caught early. Unfortunately, blood tests are poorly standardized and insensitive during the first weeks of infection. When the disease is diagnosed late in its course, it may be causing problems in many different organ systems, especially the skin, heart, nerves, and joints.

Some epidemic areas, such as eastern Long Island, Block Island, Martha's Vineyard, and Nantucket, require extra caution. Don't panic, shoppers: East Hampton's sidewalks are infested with neither mice nor deer. And keep an eye on your dog, who can get Lyme disease too (though cats seem to be less susceptible, perhaps because they clean themselves so fastidiously).

• *Brucella* •

Most of us think of infectious disease as something caught from another person, but there are a few that can be picked up directly from animals. This kind of contagion is called *zoonosis*, a word that sounds threatening but should never keep anyone away from the zoo.

The genus of bacteria *Brucella* is named after Sir David Bruce (1855–1931), a Scottish microbiologist and physician who first isolated them in 1887. (He also discovered that the tsetse fly carries the protozoal parasite *Trypanosoma brucei*, responsible for African sleeping sickness.) It has several members that mostly infect beasts with the illness now called brucellosis but can also cause trouble in people. Because Bruce and other British army scientists conducted seminal research on the island of Malta on goatherds suffering from the disease in the latter half of the nineteenth century, the disease used to be known as Malta, Mediterranean, or Gibraltar fever. In the United States, brucellosis is largely confined to slaughterhouse workers but is sometimes connected to eating a fashionable gourmet item, fresh goat cheese. Cheese made from pasteurized milk is okay.

Human brucellosis is caused by several different *Brucella* species: *B. abortus*, which resides in cattle, where it causes contagious bovine abortion; *B. suis*, which prefers hogs and is more virulent when it attacks people than its cow-rustling cousin; *B. melitensis*, which comes from goats and sheep and causes the most severe illness in humans; *B. rangiferi*, which likes reindeer and caribou; and *B. canis*, a dog lover. Many other animals, including horses, may carry *Brucella* bacteria. Humans are a dead end for these organisms, since there is no person-to-person contagion. But their zoonotic ubiquity makes them of concern as potential bioterrorist weapons.

The action of *B. abortus* in causing cattle stillbirths is a classic example of the "specificity" of infections. This organism needs certain vitamins and other substances to grow, particularly one called erythritol, which is something like sugar. It just so happens that there is lots of erythritol in calf placenta, but not much elsewhere in a pregnant cow. So *B. abortus* thrives in the exact place where it can efficiently kill the animal's embryo. Happily, human placentas do not contain erythritol.

Brucellosis is rarely fatal, but the fevers (which go up and down, thus giving the disease another nickname, undulant fever), chills, weakness, and depression it brings are debilitating. Richard Nixon suffered from undulant fever as a teenager. *B. melitensis* is infamous for causing abortions in women, especially during the first trimester. In 1918, the researcher Alice Evans warned that *B. melitensis* was very much like the germ that caused contagious abortion in cattle. She caught brucellosis in 1922 after laboratory exposure and suffered from recurring fevers for twenty years, but she gradually recovered and lived to the age of ninety-four.

In developed countries dairy herds are monitored for the bacterium's presence through tests of bulk milk. In any case, pasteurization zeros the risk of contracting brucellosis. Though vaccines are available for animals, they are too toxic for humans.

• *Campylobacter* •

For many years, bacteria of the genus *Campylobacter* were thought to be related to the *Vibrio* germs of cholera infamy, which also produce intestinal illnesses. *Campylobacter* turns out to be much less dangerous but far more widespread. It accounts for anywhere from 5 to 14 percent of all diarrhea in humans worldwide, making it the most common bacterial cause of this misery. It is one of the Big Four of the so-called enteric—that is, pertaining to the intestines—pathogens, along with *Shigella, Salmonella,* and *Giar-*

dia lamblia. Their basic effects are very much alike and require no elaboration here, please. They travel the ever efficient fecal-oral highway via contaminated food and water whenever sanitation, hygiene, or cooking are, shall we say, inattentively carried out.

Two English veterinarians with strong stomachs were the first to identify *Campylobacter* microorganisms in 1909, while studying aborted cattle fetuses; they named the bacteria *Vibrio fetus.* In 1947, the same germs were found in a woman's blood under analogous circumstances. Not until 1972 did researchers realize that they were dealing with a new genus, which they named after the Greek for "curved rod," the bacteria's characteristic shape. Most of the illness in people in the United States is associated with handling or eating raw and undercooked poultry—more than half the raw chicken in markets is probably contaminated—and is caused by one species, *C. jejuni.* In rare instances, perhaps one out of every thousand cases, campylobacteriosis leads to Guillain-Barré syndrome, a crippling and degenerative paralytic illness in which a person's own immune system attacks his or her nerves.

Darwin would be pleased to know that campylobacters just happen to thrive at the body temperatures of their warm-blooded hosts, which include cattle, horses, birds (the favorite home for *C. jejuni*), and many popular pets, as well as people. "Thrive" may actually be too timid a word, since a single gram of an infected person's stool may contain a billion campylobacters—plenty to go around.

Never eat pink chicken meat. Scrub that sloppy cutting board. And wash your hands a lot if your puppy or kitten is sick.

• *Candida albicans* •

Candida albicans is a yeast, which is a kind of fungus, and it is definitely among us. It is not the sort of yeast that makes bread rise or beer brew or cider ferment, however. And it is not the fabulous stuff called Marmite that Brits love to smear on toast. Much like its bigger brethren growing on rotten tree trunks, *Candida albicans* is found naturally where the human body harbors the kind of decaying organic matter it likes to eat—that is, in the mouth, genitals, or skin. (Don't feel insulted—even the most squeaky-clean aesthete has a lot in common with rotten tree trunks.) For the most part, the yeast becomes a problem only if our defenses weaken to the point where it grows too much, rather like a sandwich covered with mold as the preservatives wear out.

Diaper rash happens when *C. albicans* sets up shop on a baby's bottom; the immune system is not really a factor here, just the availability of a perfectly yummy environment. So-called yeast infections of the female genital tract are due to *C. albicans* finding its way there from normal homes on the skin or in the intestines, hardly ever by sexual means. A condition known as thrush—creamy white "cottage cheese" patches in the mouth or on the tongue that can be scraped off—develops when *C. albicans* proliferates in that orifice; it gained notoriety as a sign of AIDS in young adults but is found sometimes in normal children. Candidal jamborees can also appear around fingernails. Other *Candida* species, such as *C. glabrata,* sometimes join the feast. None of these infections is serious in itself, yet most of us probably spurn the notion that we may sometimes resemble rotten tree trunks or funky sandwiches.

Baker's and brewer's yeast, by the way, is *Saccharomyces cerevisiae.* Food yeast à la Marmite, developed during the shortages of World War II, is *Candida utilis.*

• chikungunya/o'nyong-nyong viruses •

The chikungunya and o'nyong-nyong viruses are two arboviruses that cause diseases in Africa and Asia similar to dengue fever.

O'nyong-nyong, which means "weakening of the joints" in the Acholi dialect of Uganda, first appeared in Uganda in 1959, when it affected millions of people south to Malawi with fever, swollen lymph nodes, and severe joint pain. It hitchhikes on *Anopheles* mosquitoes, already infamous for spreading malaria. Chikungunya, which means "that which bends up" (referring to the stooped posture of victims) in the Swahili language of the Makonde Plateau in Tanzania, is a closely related virus, possibly a genetic precursor, that is endemic in East Africa. It rides in *Aedes aegypti* mosquitoes, which also transport dengue and yellow fevers. Since the 1950s, it has been implicated in epidemics in Asian countries, too, including the Philippines. In 2009 an outbreak of more than 34,000 documented cases occurred in southern Thailand, including the jet-set tourist island of Phuket. There is no specific treatment for either chikungunya or o'nyong-nyong, so defense consists of killing or repelling the insects, sleeping under nets, and other safari-adventure movie stuff. Western physicians tend to call these viruses "novel," a perception that millions of Africans would no doubt find peculiar but not funny.

• Chinese restaurant syndrome •

Not a germ, but sort of related. Monosodium glutamate, aka MSG, the not-so-inscrutable ingredient in Chinese food (at least the kind served to Occidentals), can cause chest pain, burning sensations, and the feeling that your face is blowing up like a balloon. The effects depend on how much MSG you eat and vary greatly from person to person. Perhaps the mischievous gods created this little curse as a reminder that MSG is manufactured by *Corynebacterium glutamicum,* a cousin of the diphtheria germ.

You, too, may have suffered from substance abuse. So just say no.

• *Chlamydia* •

It sounds like something you don't want to get on you. *Chlamydia* bacteria have been bugging humankind for thousands of years, but as lifestyles have changed, the ailments they cause have shifted in significance. Scientists once thought that *Chlamydia* were viruses, because they cannot reproduce without attaching to other cells. The fact is that they possess all the biological features needed for independence, except the ability to generate their own energy. They are therefore known as "energy parasites," stealing their host's fuel to run their personal metabolism. Hostile takeover, indeed.

Three species of *Chlamydia* are of medical interest. *C. trachomatis,* the most important, is passed around sexually and causes a variety of problems in human reproductive systems. It accounts for some three to five million infections annually in the United States, representing direct and indirect costs estimated at $3.5 billion. *C. psittacii* is carried by infected birds and gives us a kind of pneumonia. *C. pneumoniae* is a common respiratory pathogen that has also been implicated in coronary heart disease. None of these infections is usually fatal unless there are serious underlying diseases of other kinds. Still, they can be nasty.

C. trachomatis—the most prevalent of all sexually transmitted organisms, including the one that causes gonorrhea—comes in two harmful strains. One produces infections of the eyes and/or genital areas. The other brings something called lymphogranuloma venereum, which has been fading away in the United States except among gay men. This disease is more common in tropical regions of Africa, Central America, and Asia, where it is of concern to visiting (that is, invading) troops.

Strain number one's attack on the eyes, called trachoma, is now rare in the United States and Europe, but it is still the leading infectious cause of blindness in hot, dry regions of the Third

World. About 500 million people are infected, resulting in 6 to 8 million cases of blindness. It was one of the earliest human diseases to be recognized as such, being described on papyrus in 1500 B.C. The name "trachoma," from a Greek word meaning "rough" (referring to the pustules), dates to A.D. 60. The disease spread out of the Holy Land with the Crusades and became known as "Egyptian" or "military" ophthalmia. Today it passes mostly among children and then to women caring for them. Clean water and fly control usually put an end to it.

Strain number two has definitely been on the rise since the 1960s. The least bothersome thing it causes is conjunctivitis, inflammation of the rims of the eyes. The most tragic is infertility. When transmitted sexually, it can trigger a bevy of urinary tract infections in men and women, pelvic inflammatory disease, and ectopic pregnancy. It may give a baby pneumonia during birth if the mother is infected. Like other genital pathogens, it is found most often among young, nonwhite, unmarried, poor people. In more than half of those who harbor this germ, no symptoms are apparent. Antibiotics can kill it, but using condoms is the best defense if you have lots of partners. Screening tests are recommended for sexually active women 25 and under, older women with multiple sex partners, and all pregnant women.

Research suggests that chronic infection with *C. pneumoniae* puts people at risk for atherosclerosis (hardening of the arteries, which causes more deaths in the United States than any other single disease) and heart attack. If this is proved conclusively, it could be a bonanza for public health, since the organism is vulnerable to antibiotics.

As for the pneumonia-like condition called psittacosis caused by *C. psittacii,* infection from casual exposure to bird droppings is exceedingly rare. All birds are potential carriers, but pets such as parrots, parakeets, cockatiels, and macaws, as well as poultry such as turkeys and ducks, are most often implicated.

Maybe you shouldn't let the park pigeons sit on your head while you feed them.

• cholera •
(Vibrio cholerae)

Cholera is one of the great social and political forces in human history. A filth disease that spreads via water and food contaminated by the *Vibrio cholerae* bacterium, cholera is now easy to prevent and treat but remains a potent scourge wherever sanitation is not or cannot be maintained. In 2007, 53 countries reported 177,973 cases and 4031 deaths to the World Health Organization. Gabriel García Márquez chose well with his title *Love in the Time of Cholera,* there being few other kinds of experience more contradictory.

Historically, *V. cholerae*'s favorite breeding ground has been the Indian subcontinent, where clean water is still a pipe dream. Twenty thousand pilgrims were killed by cholera at Hurdwar, a holy place in northern India, in 1784. Scientists count six distinct pandemics since 1817, during which the disease raged out of Bengal—probably, at first, because of expanded trade and the movement of British troops—and attacked ever more distant populations outside the country, only to slither back into India. During the second pandemic, which reached England in the early 1830s, John Snow began his classic investigation of the Broad Street pump, a public source of water drawn from a well near a cesspool in London's squalid Soho district. Snow, an unfashionable general practitioner, believed that cholera was waterborne, carried by the excreta of victims—in those days a brilliant but hard-to-prove guess. After he persuaded authorities to remove the pump's handle, thus shutting off the water supply, the neighborhood's outbreak waned. (In charge of cholera patients

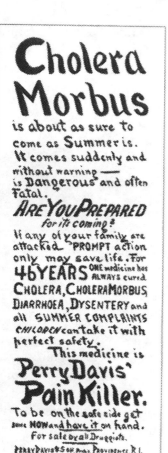

Cholera Morbus is about as sure to come as Summer is. It comes suddenly and without warning — is Dangerous and often Fatal.

ARE YOU PREPARED for its coming?

If any of your family are attacked, PROMPT action only may save life. For **46 YEARS** ONE medicine has ALWAYS cured CHOLERA, CHOLERA MORBUS, DIARRHOEA, DYSENTERY and all SUMMER COMPLAINTS, CHILDREN can take it with perfect safety.

This medicine is **Perry Davis' Pain Killer.**

To be on the safe side get some NOW and have it on hand. For sale by all Druggists.

PERRY DAVIS & SON, Man. Providence, R.I.

at the local hospital was an indefatigable thirty-three-year-old nurse named Florence Nightingale.) Snow was lucky not to be attacked by mobs who thought doctors were robbing the graves of cholera victims for anatomy lessons. On the Continent, French workers rioted because they feared they were being intentionally poisoned. Parisian ragpickers went on a rampage when the government ordered garbage collection as a preventive measure.

The disease returned to kill at least 61,000 people in Britain in 1848–49. In Paris, the same pandemic claimed 18,000 lives out of a population of 785,000, but riots by the working classes were averted when police relaxed controls such as *cordons sanitaires* around poor neighborhoods. During this era, widespread death spawned some public health innovations but more often supported campaigns for moral reform.

The fifth pandemic, starting in 1881, brought rival German and French scientists to Egypt in order to discover the pathogen. The French team mistook ordinary blood platelets for cholera bacilli, then relaxed with swims in the sea, where one must have been a gulper. As the Frenchman later lay dying of cholera, the great German biologist Robert Koch, who had realized the

French error and would soon identify *Vibrio cholerae,* comforted him by saying that he had indeed found the right germ.

All cases today are part of a seventh pandemic that started in Indonesia in 1961 with a new strain: *V. cholerae* 01, biotype El Tor. Due to the ubiquity of fast international travel and commerce, it is unlikely that the beast will ever retreat again to its Indian lair. Indeed, cholera can jump continents in a matter of days. Moreover, the El Tor vibrio can travel around the oceans in algae. Start-

An early-nineteenth-century German illustration of a man wearing protective gear against cholera—face mask, oversize shoes, a steaming pot on his head—as he pulls a wagon of his earthly possessions.

An illustration published in London in 1832 showing two physicians bundling up a cholera patient for removal while others inspect his dirty trousers and chamber pot.

ing in several coastal cities of Peru after a Chinese ship dumped contaminated bilgewater, cholera reached epidemic proportions in South America in 1991, marking the first such outbreak there in a century. Victims soon appeared in Central America and Mexico. Over three years the disease struck at least 900,000 people and took more than 8,000 lives. Another newly identified strain of unusual virulence, *V. cholerae* 0139, swept across India and Bangladesh into Thailand in 1993, infecting hundreds of thousands of people and killing thousands. A case of 0139 infection popped up in California that same year in someone who had recently visited India.

Cholera in North America is rare and isolated, most often occurring along the Gulf Coast, where raw oysters and undercooked crabs may transport endemic bacteria into unsuspecting

Epidemic in Marseilles, 1865. When nothing else works, build a big fire and dance around it like crazy.

gourmands. Between 1995 and 1999 there were 53 confirmed cases in the United States, including one death, all caused by the 01 strain. Most of the infections were acquired in other countries, but four came from eating Gulf Coast seafood. There have also been freakish imports from abroad, as in the episode of February 1992, when at least 31 of 356 passengers on a flight from Buenos Aires to Los Angeles were hit after eating contaminated seafood salad. In 1991, three Maryland revelers swallowed the bugs in frozen coconut milk from Thailand. Beware ceviche and piña coladas, gringos.

One of the most grisly epidemics in modern times struck along the Rwanda-Zaire border in the summer of 1994. Earlier that year, famine-ravaged Somalia experienced a smaller outbreak believed to have spread from Ethiopia, whose government had suppressed reports of the disease. The chaotic circumstances of refugee camps create perfect conditions for spreading cholera through fecal-oral contagion. Humans are the only carriers of *V. cholerae,* so when hundreds of thousands of people crowd together in a prevalent region with no toilets or fresh water, the bacteria are virtually certain to start rampaging.

Yet, contrary to the hellish images from Rwanda, most cases of cholera around the world are mild or unnoticeable. Unless there is a gross source of microbes, like contaminated water or the clothes of victims, the disease is relatively hard to catch. Chlorinating drinking water prevents it. Salivary and stomach acids are a natural defense against *V. cholerae* until overwhelmed by sheer numbers of bacteria. Three-quarters of all infections show no symptoms. But severe forms—aptly known for centuries as cholera morbus—in which the victim has ingested millions of bacilli by drinking badly polluted water, for example, can be dramatically fatal when untreated. The bacteria multiply in the small intestine and produce a toxin that allows them to penetrate the protective mucous layer there, triggering a horrific outflow of bodily fluids and salts by the gallon. Sudden stomach cramps are followed by violent diarrhea, vomiting, and fever while ruptured capillaries turn the skin black and blue (whence comes the French phrase for transfixing fear, *une peur bleue*). Radical dehydration can bring death within a day through collapse of the circulatory system. Such gruesome effects contribute to cholera's fabled psychological impact on threatened populations.

If the victim's fluids and electrolytes are replaced in time, however, recovery is just as swift. Treatment could hardly be simpler: drinking, or taking by intravenous drip, large volumes of a solution of water, salts, and sugar. Ten thousand cases of

Gatorade were shipped to the Rwandans as a quick approxima-
tion. UNICEF and other relief agencies maintain supplies of jiffy
packets of salts and sugar for mixing with clean water. Antibiotics may be used to shorten the period required for absorbing
fluids and to ward off infection among the patient's immediate
household, but they are not distributed widely during epidemics
because of the likelihood that resistive *Vibrio* strains would soon
evolve. The Rwandan strain was determined early on to be
resistant to tetracycline, a standard anti-cholera drug. The horror of cholera in refugee situations is thus compounded by the
knowledge of how many lives—especially those of children,
who may account for the most fatalities—could easily be saved.

For obvious reasons, cholera has traditionally been a disease
of poor people living in sordid conditions. An exception was
the 1970 outbreak in Jerusalem that struck the upper classes,
many of whose victims were found to have highly alkaline
stomach contents, which acid-hating *V. cholerae* could pass
through unscathed. Terrible epidemics that struck New York,
Boston, Philadelphia, and other American cities in the nineteenth century stigmatized slum-dwelling immigrants as the
source of infection through their perceived intemperance and
lack of hygiene. Previously unknown in the Western Hemisphere, cholera was considered alien to the bucolic New World
experience—more the burden of poor Eastern Europe, say,
where Poles cursed by saying "Cholera!"—meaning "Damn!"
From immigrant neighborhoods it spread across the lowest strata
of urban society, forming an indelible relationship between disease and vice. The young nation's reaction to related issues of
poverty, deviance from the mainstream, and socioeconomic
assimilation was played out again on the AIDS front.

Cholera often serves as an alarm signaling a public health system under stress. In 1994, an epidemic erupted in southern Russia and spread across Crimea into Ukraine. Former Soviet officials
once proud of what had been accomplished under socialism

were faced with a poorly funded network of surveillance and treatment exacerbated by industrial pollution of water supplies.

Other members of the *Vibrio* genus may cause lesser diarrheal illness, blood poisoning, and wound infections—but none is as monstrous as *V. cholerae.* For example, *V. parahemolyticus* accounts for the majority of seafood poisoning cases in Japan during the summertime (usually from shrimp). It has also been reported in coastal American states such as Maryland. *V. alginolyticus* and *V. vulnificus* can infect wounds received in warm saltwater environments. The latter germ can also cause blood poisoning if swallowed, most often in people with weak immune systems or liver disease. *V. vulnificus* has turned up in raw Gulf Coast oysters, causing at least thirteen deaths since 1992. As the name suggests, all vibrios have a little tail that makes them frenetic swimmers (Koch compared them to swarming gnats), a feature that resonates with their penchant for ruining good water.

An oral vaccine exists but it confers only brief and incomplete immunity. It is not available in the United States and not recommended for most travelers.

Kenne hora, as the Yiddish blessing goes: No cholera!

• *Clostridium* •

The genus *Clostridium* is a group of bacteria that, like large human families, has good and bad members. Named for the Greek word that means "spindle," these bacteria, which are cowardly themselves, produce rugged spores that can be found in dirt and dust all over the planet. Some are natural inhabitants of the intestinal tract, and thus the excreta, of many different mammals, including cattle, horses, and humans. Others are supreme opportunists, setting up house in a damaged host, where they produce poisonous toxins as they multiply. Because they are anaerobic and cannot thrive in fresh air, they love rotty places or sealed-off spaces. A few are therefore wicked causes of food poisoning and wound infections.

Clostridium botulinum is perhaps the family's blackest sheep; it is actually a gang itself, consisting of seven members, of which four may harm us. About two hundred years ago, scattered cases of muscle weakness and respiratory failure were linked to eating sausage and fish in Germany, Scandinavia, and Russia. The condition, which could quickly lead to death, became known as botulism, from the Latin word for sausage, *botulus*. In 1897, a researcher named Van Ermengem found in a culprit ham a microscopic organism that would multiply only in oxygen-free containers. He went on to determine that it was not a parasite; that is, botulism was not an invasive, infectious disease but the result of eating an incredibly powerful poison left in the ham by the microbe. This toxin could be destroyed with heat, he noted, and the organism would not produce it at all if the food was too salty.

Incidents of botulism rose with the practice of home canning, reaching a peak in the United States of about thirty-five deaths per year in the 1930s—very few, but almost always the subject of publicity. The microbes release a gas, and swollen tin cans became the stuff of household nightmares. Thanks to refrigeration and modern preservation techniques, foodborne botulism is unusual today. Out of an average of 145 cases reported every year, only about 15 percent are from eating something bad. The largest twentieth-century outbreak in the United States occurred in 1977, when fifty-nine people were stricken after eating poorly preserved jalapeño peppers. In 1993, two restaurant outbreaks in Georgia and Texas were connected to canned cheese sauce and a dip containing baked potatoes, respectively.

Because *C. botulinum* spores are so ubiquitous in nature, it is conceivable that botulism could happen as an infection—in a dirty wound, say, or in the intestines—rather than from swallowing toxin already formed in spoiled food. In fact, this turns out to be extremely uncommon in adults; yet because infants' intestines are vulnerable to dust containing spores, botulism may take this path in children under about eight months of age (accounting for 65 percent of cases each year in the United States). Spores in honey have been implicated in some of these cases, so it is not recommended for babies. Wound botulism has been on the rise because of the use of black-tar heroin by injection drug users, especially in California. The toxin is so potent that even a lethal dose is too small to trigger our defensive system to produce an antidote, so immunity does not follow such poisoning. If untreated, botulism can show a 65 percent fatality rate. Attentive medical care brings this down to less than 10 percent. No licensed vaccine is currently available for public use.

The tenacious spores of *C. botulinum* must be heated above 250°F (which is possible only in a pressure cooker) to guarantee destruction. They hate nitrites—hence the presence of that additive in hot dogs and other semicooked meat products. All in all, botulism is quite unusual nowadays, but anyone who "puts up" the bounty of a vegetable garden should be careful to follow home economics instructions. A supply of antitoxin against botulism is maintained by the Centers for Disease Control and Prevention. An experimental botulinum toxoid vaccine was given to about eight thousand military personnel during the Gulf War.

Because of its many nasty attributes, *C. botulinum* toxin is listed among the "Category A," or most dangerous, bioterrorist agents. Still, anyone who is unable to cope with the realities of lost youth can call upon it for at least a temporary cure. For about $350, carriage-trade dermatologists will inject the stuff into eyebrow muscles, thus paralyzing them and smoothing away worry wrinkles. The treatment guarantees three to six months of looking blissful as a debutante. Because a tube of the toxin loses its strength a few hours after being opened, Park Avenue doctors hold "botulism days" for all their eager patients. Enjoy.

C. perfringens is a more common, though far less hazardous, cause of food poisoning in the United States, especially in cooked beef and poultry. Ubiquitous in nature, its spores may survive the initial high temperatures of cooking, then germinate as the food cools and multiply if not properly refrigerated. If the food is served without enough reheating, live bacteria are then eaten by hungry human hosts, who get cramps and diarrhea eight to sixteen hours later. In 1993, two sizable outbreaks in Ohio and Virginia were traced to bad corned beef on St. Patrick's Day.

Far more threatening is the dirty work of another family member, *Clostridium tetani,* the tetanus bacterium. It, too, is dangerous because of the nasty toxin it releases, but not if you swallow some. Spores in soil may enter the body through almost any kind of wound—not just deep ones like the classic rusty-nail puncture, but even trivial splinter cuts—where they are then activated by decomposing tissue. Their toxin quickly finds its way through the nervous system into the spinal cord, making muscles contract (thus the nickname "lockjaw") and triggering spasms. A hideous grin, called *risus sardonicus,* may disfigure the face of untreated victims. Happily, the disease is totally preventable and now rare in developed countries, except among intravenous-drug abusers. Fewer than fifty cases are reported in the United States every year, but the fatality rate is 20 percent.

The peril of tetanus has been recognized at least since the time of Hippocrates in the fifth century B.C., and certainly in ancient Egypt, where dung was a favored wound medicine. It was especially notorious during military campaigns, when filthy lacerations and burns were the rule. After the Battle of Waterloo, the Scottish anatomist Sir Charles Bell wrote that "there are but two cases of living in all that had this affliction." Rather like botulism, tetanus was not understood until the late nineteenth century, but even then it was of major importance only on the battlefield. Later, mushrooming numbers of automobile accidents widened its swath. The great Japanese bacteriologist Shibasaburo Kitasato (1852–1931) showed that the damage was caused by the microbe's toxin. In order to find out how fast the poison traveled, he injected bacteria into the ends of mouse tails and then chopped the tails off as soon as an hour later, but too late for the mice.

Perhaps the most important discovery was that *C. tetani* toxin could be treated with chemicals like formaldehyde and made nonpoisonous—such a substance is called a *toxoid*—without ruining its ability to tweak our immune system into manufacturing a natural antitoxin. The toxoid was first used successfully on French cavalry horses after World War I. In addition, small amounts of toxin could be injected into animals, whose systems then produced an antitoxin that can be given to humans in emergencies (it does not bring the patient permanent immunity). Horses, especially sensitive to tetanus, are used for commercial production of the antitoxin. Thus, two routes of preemptive defense were established that have virtually wiped out the disease: long-term immunization with tetanus toxoid and/or injection with tetanus antitoxin within a day or two after injury. The first is nearly foolproof, the second not. Between 1982 and 1998, 63 percent of tetanus cases in the United States occurred in people over fifty years old, an age group unlikely to have up-to-date immunization.

One heartbreaking indication of the enormous public health gap between developed nations and the Third World is the existence of so-called neonatal tetanus. Hundreds of thousands of babies die every year in Asia and Africa because their umbilical cords are cut with filthy instruments or the stump is packed with dirt.

Clostridium difficile can sometimes cause colitis in those taking antibiotics. About 3 percent of healthy adults carry it around as part of the normal "flora" in their intestines, but it can also be picked up through fecal-oral contact, often in hospitals or nursing homes. Antibiotics change the natural balance of bacteria in the gastrointestinal tract, allowing *C. difficile* to multiply to the point where its toxin causes diarrhea problems.

At least one kind word can be said about clostridia. *C. pasteurianum* is one of the bacteria in agricultural soils that help plants acquire nitrogen. It is thus fundamental to the world's food production and perhaps more than makes up for the havoc wreaked by its fiendish relatives.

• Congo-Crimean/Rift Valley/California/
St. Louis viruses •

These far-flung locations are mainline stations on the arbovirus express. Different viruses may cause diseases with symptoms that are clinically indistinguishable. They can be told apart by delving into the body's biochemistry, but for the most part they are known for their geographical loyalties. Thus the melodiously malevolent chorus of locales associated with encephalitis and sundry fevers that in this case are carried to humans by insects.

A hemorrhagic fever—one accompanied by rash, bleeding gums and other mucous membranes, and sometimes internal bleeding—with high mortality (15 to 20 percent) was first reported in rural areas of the southern Soviet Union during World War II. It appeared to come from the bite of ticks that infest rabbits. The culprit virus, technically in a category of arboviruses called *nairoviruses,* was isolated in the mid-1960s and named Crimean, but researchers in Africa found the same germ and called it Congo. So the scientific honor was split between the two regions, which would probably rather do without such an accolade. In Africa, the virus has been found in all sorts of creatures besides ticks—cattle, goats, hedgehogs, and gnats—but very few humans.

Rift Valley fever traditionally haunts the lands south of the Sahara Desert, bringing serious problems to people and livestock. Caused by an arbovirus of the Bunyaviridae family, transmitted by *Aedes* and other kinds of mosquitoes, it was first noticed in 1930 as producing stillbirths in European breeds of cattle in Kenya and fever in veterinarians who investigated the problem and were thereby exposed to the animals' blood or viscera. After the completion of Egypt's Aswan Dam in the 1970s, which created vast new wetlands conducive to mosquito breeding, an epidemic hit some 18,000 people in the region and killed 598, also wiping out livestock herds. Epidemics related to dam construction happened in Mauritania, Senegal, and Madagascar during the next decade. Livestock have been extensively vaccinated in Egypt. In 2000 Rift Valley fever caused 843 cases with 114 deaths in Saudi Arabia and 1087 suspected cases with 121 deaths in Yemen, probably through importation of infected livestock. A vaccine for humans, called RVF MP-12, is still being tested and thus unavailable.

California and St. Louis are two of the five arboviruses of greatest significance to humans in the United States. They both cause encephalitis and are carried by mosquitoes (the former mainly by *Aedes triseriatus,* which breeds in tree holes and other containers, and the latter by several *Culex* types, which are the most common night-flying mosquitoes in North America). Neither is confined to its namesake location. California encephalitis is seen most often in forest workers or other people, like campers, who spend lots of time in wooded areas. St. Louis encephalitis happens in irrigated farm country and urban or suburban areas where standing water provides breeding opportunities for *Culex* biters.

The other three important arboviruses cause eastern equine encephalitis, which is often fatal; western equine encephalitis; and Colorado tick fever, which is transmitted by the wood tick

Dermacentor andersoni. Though unpredictable outbreaks occur from time to time, none of these diseases is common.

• coronaviruses •

Coronaviruses are some of the little devils behind the common colds that afflict people of all ages worldwide—and are thus one of the inescapable annoyances of human life on earth. Named for their crown of club-shaped thorns, which can be seen with an electron microscope, they cause colds especially during the winter and spring (when another major group of runny-nose pests, the rhinoviruses, is at a low ebb).

They are not all trivial. Between November 2002 and July 2003, a previously unrecognized coronavirus caused at least 8098 cases of a far more serious illness called severe acute respiratory syndrome (SARS) in 29 countries, with 774 deaths. First reported in China, where it probably jumped from bats through market animals to humans, the SARS-CoV virus spread worldwide before being contained after the WHO issued a global alert in March 2003. In the United States, there were only eight confirmed cases—all suffered by people who had traveled through SARS zones abroad—and no deaths, but considerable anti-Asian paranoia arose. Generating scary headlines at first, SARS turned out to be a relatively rare disease that could nonetheless re-emerge at a future point, depending on the vagaries of nature.

Coronaviruses need only about three days to multiply in the respiratory tract before their victim starts feeling miserable. On average, the cold lasts for a week—a few days shorter than a typical rhinovirus cold, but with more nasal congestion. Coronaviruses are remarkably good at reinfecting their hosts, which is one reason why vaccines that protect against them remain elusive.

This germ is important not because of the severity of the disease it causes but because of the sheer number of clogged-up, grumpy humans it endlessly afflicts. What evolutionary purpose does it serve? Why does time only move forward?

• *Cryptosporidium* •

Americans are not used to getting sick from their tap water, but parasites of the genus *Cryptosporidium*—specifically the protozoan *Cryptosporidium parvum*—can cause trouble that used to be seen mainly in poor countries with unreliable water treatment or none at all.

Cryptosporidium organisms are about one sixtieth the size of the dust particles you can watch floating in a sunbeam. They have been found in 65 to 87 percent of surface water samples tested all over the United States. Egg-like *Cryptosporidium* particles called "oocysts" are passed in the feces of infected animals such as cattle and then find their way through physical contact or contaminated water into human intestines, where they hatch an infective spore that starts another reproductive cycle. Meanwhile, the person who swallows enough of them gets watery diarrhea and cramps for about two weeks, becoming highly contagious to others via the same route. Otherwise healthy people will soon cure themselves, but anybody with an impaired immune system is in for serious setbacks. There is no effective treatment besides replacing lost fluids.

Like all waterborne infectious diseases, cryptosporidiosis becomes a mass health hazard when water-treatment plants fail to do their job. In April 1993, some 300,000 Milwaukee residents developed severe diarrhea because city water supply officials did not correct increased turbidity—that is, cloudiness of the drinking water that could be due to sand, silt, other nonorganic matter, or harmful organisms—during the spring runoff season. The outbreak, which led to 4,400 hospitalizations, was the largest ever in an industrialized country.

In December 1993, more than a million residents of Washington, D.C., and northern Virginia were advised to boil drinking water after treatment workers at two U.S. Army Corps of Engineers plants bungled the operation of filtration systems that draw water from the Potomac River. The public did not learn about the health hazard until the day after it started. Luckily, no illness was traced to this mishap.

Because cryptosporidia are not killed by routine chlorination (lab tests showed that the Milwaukee bug could live in Clorox), boiling water is the only practical defense against suspected contamination. Federal agencies recommend a rolling boil for one minute to wipe out all major waterborne protozoal and bacterial pathogens (this will also clear away hepatitis A, one of the more heat-resistant viruses that can spoil a good drink). If you live at an altitude above 6,562 feet, boil for three minutes. Cryptosporidia have also been found in oysters contaminated by sewage or runoff from pastures—another sad reason to think twice about eating these delicacies raw. And don't gulp the swimming pool water! Especially in the kiddie area.

• cytomegaloviruses •

Cytomegaloviruses, so-called because they cause enlargement of the cells (*cyto-* means "cell"), are catchall, do-all, everywhere, everybody viruses that are almost as much a part of human life as carbon and oxygen. Ninety percent of adults have probably been hit by these things, most without any symptoms, but a small minority contract liver disease, a mononucleosis-like syndrome, or respiratory problems. As is the case for many viruses, the reasons they make some people sick and others not are unknown. They can sometimes be very damaging to newborns if the mother is infected for the first time during pregnancy—in countries where routine vaccination has defeated the rubella (German measles) virus as a cause of birth defects, cytomegalovirus infections may be the major viral source of such problems. More than 80 percent of toddlers—especially kids in day care—pick them up harmlessly as they pass around in saliva, tears, urine, and feces. People with unhealthy immune systems, particularly AIDS patients, may be direly affected, however. As members of the herpes group of viruses, they show the sneaky ability to reactivate after long periods of latency that so confounds attempts to fight them when necessary.

The cytomegalovirus particle is one of the biggest of all animal viruses, but at 200 billionths of a meter (200×10^{-9} meters, or 200 nanometers) in diameter it is still way beyond the reach of optical microscopes. (The single-celled paramecium that grade-school biology students often get to watch is more than a thousand times larger than a cytomegalovirus.) It is marvelously adapted for living in human beings and hates to be anywhere else. Years of searching for an infectious agent in certain cases of infant death led to isolation of the virus in 1956 from salivary gland tissue (hence the medical nickname "salivary gland virus").

Cytomegaloviruses often reside in blood vessels during infection, raising the fascinating possibility that they play a role in the cell changes that precede hardening of the arteries, or atherosclerosis. Perhaps far in the future we'll be vaccinated against heart attacks.

• **dengue** •

Some scholars maintain that "dengue" is a Spanish adaptation of the Swahili *Ki denga pepo,* used to describe a cramp that seemed to seize people like a spirit. Others point out that dengue is also a Spanish word meaning "affectation," an unkind reference to the mincing walk adopted by untreated victims of the disease to lessen the hellish pain in their joints. The English thus named it "dandy fever." Dr. Benjamin Rush (1745–1813), the most eminent American physician of his day (and a signer of the Declaration of Independence), called it "break-bone fever" when it appeared in Philadelphia in 1780, because that's what it felt like. Today dengue occurs in the United States only when brought by travelers from areas where it is endemic, like the Caribbean islands, Central and South America, Africa, or Southeast Asia. In these locales, where billions of people are at risk, dengue is on a par with malaria among insect-borne diseases.

Dengue fever is caused by an arbovirus transmitted mainly by bites of *Aedes aegypti* mosquitoes (which can also carry yellow fever and viral encephalitis), a variety of the common summertime pest that likes to breed in containers of still, clean water. The virus jumped from monkeys to humans in Africa or Southeast Asia and probably spread with the slave trade and other commerce. While firmly established in Texas and Florida, American *A. aegypti* are not generally dangerous, because circles of dengue infection involving densely populated human settlements do not exist here anymore. An epidemic of more than a million cases occurred in the United States in 1922, but the last endemic outbreak of dengue fever happened in 2005 along the Mexican border of south Texas. Only a dozen or so confirmed cases appear every year in the United States. Local outbreaks can occur at any time if someone brings the virus in from abroad and *A. aegypti* cooperates by biting the infected person, incubating the virus, and then biting another person. For the most part, though, dengue is found today only where governments either can't afford or don't care about mosquito control. Scattered cases of dengue began to appear among thousands of American soldiers returning from Haiti in late

1994, raising concern that the virus could be introduced to wider populations—but it wasn't.

Dengue is known for bringing on a sudden 104° temperature, nausea, vomiting, a horrendous headache, a rash that appears after a twenty-four-hour pause in the fever, and a long convalescence, but it is not deadly. Most cases are mild, treated with fluids and bed rest. No vaccine is currently available.

There is a far more serious form of the disease, called dengue hemorrhagic fever, whose incidence has been increasing since it was first reported from Thailand and the Philippines in the mid-1950s. It hit 10,367 people in Bangkok between 1958 and 1963, killing 694. The dengue arbovirus comes in four models—they're all basically Chevrolets, so to speak, with different trim—and if you pick up two or more of them simultaneously, you get the really bad kind of dengue. In 1981, Havana was overwhelmed by an epidemic of dengue hemorrhagic fever that struck 344,000 out of a population of 2 million and killed 158. In 1982, more than 20 percent of New Delhi's 5.6 million people caught the disease. More than 3,100 cases occurred in Venezuela in 1990, with 73 deaths. The four types were probably once confined to separate geographic areas, but worldwide travel by infected people has mixed them together.

Another *Aedes* mosquito—*A. albopictus,* the "Asian tiger"—can also transport the dengue microbe. It arrived in Houston, Texas, for free with a shipment from Japan of used tires—a favorite breeding place—in 1985 and is now established in at least seventeen states. Though the aggressive tiger has not yet been found to carry dengue in the continental United States, another dangerous arbovirus, eastern equine encephalitis, has been found in *A. albopictus* around a tire dump in Florida.

• diphtheria •
(Corynebacterium diphtheriae)

George Washington died in his bed at his Mount Vernon plantation on December 15, 1799, because his throat was so swollen he couldn't breathe. It was a ghastly end for the Virginia gentleman, perhaps the result of infection with *Corynebacterium diphtheriae.* (The genus is so named because all its members are club-shaped, from the Greek *koryne,* meaning club.) Though historians still debate whether George was done in by diphtheria, the disease's status in literature is ensured by William Carlos Williams's 1933 story, "The Use of Force."

Although outbreaks of such deadly sore throats had been noted sporadically since the time of Hippocrates, epidemics seem to have become common beginning in the sixteenth century. To prevent suffocation, tracheotomies were first used widely during an epidemic in Naples in 1610. In 1659, Cotton Mather described a "Malady of Bladders in the Windpipe" in Massachusetts that was especially vicious in children. Another New England epidemic started in 1735, taking about a thousand lives, of which nine hundred were children. In 1826, during a period when the disease was prevalent in France, the physician Pierre-Fidèle Bretonneau (1778–1862) aptly named it after the Greek word for a piece of leather, *diphthera,* referring to the tough, gray, mucousy membrane of dead cells that may block the victim's throat. The Spanish had once used an even more graphic term, calling it *garrotillo* after the executioner's implement, a string around the neck tightened by twisting a stick.

By the mid-nineteenth century the microbe responsible for diphtheria apparently became even more potent, because severe cases began to multiply sharply worldwide. In New York, where annual deaths from diphtheria averaged about 325 in 1872, fatalities jumped to more than 2,300 in 1875. *Corynebacterium diphtheriae* was identified in 1883, and by 1890 it was known

CARRIER OF

DIPHTHERIA

KEEP OUT OF THIS HOUSE   By Order of BOARD OF HEALTH

Any person removing this card without authority is liable to prosecution

Board of Health quarantine notice, San Francisco, circa 1910.

that the bacterium produced a poison, called an exotoxin, which could be used in weakened form to prod the human immune system into producing a defensive material, called antitoxin. In this regard it was similar to tetanus (see *Clostridium*), whose mysteries were being unraveled at about the same time. Routine immunizations began in the 1920s, when as many as 200,000 cases and 15,000 deaths were occurring annually in the United States, leading to near eradication of the disease wherever such programs are maintained. Not until 1951, however, was perhaps the most curious microbiological fact about diphtheria discovered: *C. diphtheriae* microbes are themselves infected with viral parasites called bacteriophages. Actually, the gene necessary for producing diphtheria toxin comes from the bacteriophage, not from *C. diphtheriae*. Without it, *C. diphtheriae* would not be virulent.

Today, the diphtheria-tetanus-pertussis (DTaP) shot is among the first that children receive after birth. Between 1980 and 1999, only 49 cases of diphtheria were reported in the United States, mostly among immigrants and/or due to lax immunization. By contrast, with public health systems under severe stress, a major epidemic began in countries of the former Soviet Union in 1990, causing some 150,000 reported cases and 5,000 deaths by the end of 1996.

C. diphtheriae bacteria thrive in dark, wet places like our upper respiratory tracts. Contagion usually occurs through close-up coughing or hand to mouth contact. Crowded, unhygienic living conditions help the microbes spread from person to person. Once established in a victim's tonsils, say, *C. diphtheriae* microbes are among the fastest microorganisms to produce symptoms of illness—not by racing through the body, but by the formation of a powerful exotoxin. The poison may spread to damage tissues in the heart, kidneys, or nervous system. Death often comes from inflammation of the heart. Unimmunized sufferers must be given a quick jolt of antitoxin plus antibiotics and be isolated under intensive care.

One of *Corynebacterium diphtheriae*'s cousins, *C. glutamicum,* has a much friendlier relationship with the human race. It is used to manufacture the all-purpose flavor enhancer monosodium glutamate, or MSG.

• dysentery •

"Dysentery" is a polite word coined by Hippocrates for the bloody runs. Its miseries have been described since the beginning of recorded history—Horus, son of the Egyptian gods Osiris and Isis, had dysentery. It has been of incalculable importance in military matters: From the Peloponnesian War through Napoleon's campaigns to the great sieges of the twentieth century, "campaign fever" has claimed more victims through contaminated food and water than has combat. William the Conqueror knew it well. Edward I and Henry V of England both succumbed to it, and the French mocked the dysenteric (though successful) English invaders at Crécy in 1346 as "barebottomed." George Washington suffered recurrent episodes of dysentery, exacerbated by hemorrhoids that sometimes required him to ride with pillows on his saddle. During the Civil War, the Union Army lost 81,360 soldiers to dysentery (and typhoid, with which it was then grouped), compared with 93,443 killed in combat. It can be caused by *Shigella* bacteria (the genus named after Japanese bacteriologist Kiyoshi Shigu, who discovered *Shigella dysenteriae* in 1898), in which case it is more properly known as shigellosis; or an amoeba, *Entamoeba histolytica,* a slightly higher form of life, whereby it is called amebiasis. Human excrement is the primary source of infection, with amebic dysentery prevalent where night soil is used as fertilizer. *E. histolytica* may travel through the bloodstream in prolonged cases to cause dangerous abscesses in the liver and brain.

Dysentery today appears sporadically where living standards are high: in overdeveloped suburbs where wells have been dug too close to septic systems; in day-care centers and preschools where kids do not wash their paws between potty and table; or aboard cruise ships where the buffet is more dazzling than the galley deck. *Shigella sonnei* accounts for about two-thirds of shigellosis in the United States, and *Shigella flexneri* for almost all the rest. About 18,000 cases a year are reported in the United States, but the actual number is probably twenty times higher. In the Third World, where public sanitation is lax, *S. dysenteriae* still causes deadly epidemics in wide populations. Since the 1960s, drug-resistant strains of *S. dysenteriae* have been cropping up around the planet, sharply increasing the fatality rates for dysentery in poor countries. Lack of effective antibiotics is especially dangerous for children, one-fifth of whom may die if they are untreated for *S. dysenteriae* infection (four-fifths of children living in the tropics acquire bacillary dysentery before the age of five). In Ontario, Canada, 50 percent of all *Shigella* cases in 1990 were the result of organisms resistant to at least four antibiotics. Attention, adventure trekkers: Self-medicating with antidiarrheal agents such as Lomotil or Imodium can make a bad shigellosis situation worse.

New *Shigella* strains have appeared in fast-food hamburgers, which may not get fried enough to kill the pests. The bacteria, and thus the burgers, are uniquely potent, given that fewer than two hundred microbes can easily produce the disease in healthy adults. Yet the bacteria rarely cause severe illness in otherwise healthy babies, probably because breast milk contains defensive agents. It seems unlikely that fast-food restaurants will somehow incorporate this knowledge into their menus.

• *E. coli* •

The rod-shaped bacteria known as *E. coli* (the *E.* stands for *Escherichia*) are part of the normal "flora" of our intestines. The fact that they come in many different strains and are capable of reproducing at astronomical rates—one *E. coli* cell could multiply into a mass greater than the earth in three days if enough food were available—is the basis of their scientific usefulness and potential danger.

The vast majority of *E. coli* strains are harmless to most humans most of the time. For years they have served as the workhorses of biotechnology, providing fast-forward evolution for myriad types of experiments. However, the fact that they can multiply so quickly, doubling their population every two hours under the right conditions, means that they have the potential to make people, especially children and the elderly, sick. There are four major groups of *E. coli* whose toxins cause diarrhea of various levels of severity. Some are the culprits behind traveler's tummy or "Montezuma's revenge." One country's flora are different from another's, so tourists may suffer a kind of poisoning until their immune systems adapt to newly imbibed bacteria. Others primarily infect Third World children. Still others are a problem for immunocompromised persons such as those with AIDS.

The worst *E. coli* bacterium, *E. coli* 0157:H, a mutation discovered in 1982 that has at least sixty-two subtypes, causes hemorrhagic colitis, an acute bloody diarrhea much like dysentery that can kill weak victims. Undercooked beef and raw milk are the microbe's favorite homes. A single outbreak in Wishaw, Scotland, in 1998 made five hundred people sick and caused twenty deaths. Some seventy-three thousand cases of infection and sixty deaths occur annually in the United States. ECO157, as it is abbreviated, is also the leading cause of a kidney disease called hemolytic uremic syndrome. Like many other strains of *E. coli,* this one's resistance to drugs is apparently due to overuse of antibiotics in livestock—it may live in the intestines of healthy cattle and then contaminate meat during slaughter. In 1991, twenty-seven people in Massachusetts were struck with ECO157 infection after drinking cider made from apples grown on trees fertilized with manure. In January 1993, contaminated fast-food hamburgers from the Jack-in-the-Box franchise triggered a multistate outbreak that killed at least four children. It struck not only those who ate the burgers but some who ate salads made with the same utensils used on the meat, as well as others who merely came into contact with people already infected. More than five hundred in the Seattle area got sick, with kidney failure occurring in at least twenty-nine. In 1994, the U.S. Department of Agriculture declared that raw ground beef found to be contaminated with ECO157 would be considered "adulterated" and prohibited from sale—the first time such drastic action was ever taken on raw food just for containing pathogenic bacteria. But this still has not prevented tens of thousands of people from being sickened annually. Consumers can protect themselves at home by cooking ground beef to 150°F or until the inside is not pink and juices run clear.

ECO157 can also be waterborne, a fact discovered the hard way in July 1991, when eighty people were infected while swimming in an Oregon lake. In July 1993, thirty-five thousand

New Yorkers had to boil their water because ECO157 turned up despite chlorination.

Most *E. coli* are spread by contaminated food and water. So drink the beer and ask for "well-done."

• *Ehrlichia* •

Bacteria of the genus *Ehrlichia* prefer to invade white blood cells, the front line of human defense against harmful microorganisms. They are carried into humans by Ixodidae, or hardbacked, ticks that have fed on the blood of infected mammals such as deer, elk, rodents, and perhaps dogs. Unlike Lyme disease and Rocky Mountain spotted fever, two tick-bite diseases that occur mostly in children, ehrlichiosis is most common in people over forty. Many infections may go unnoticed, but those who do fall ill experience fever, aches, nausea, and, in severe cases, prevalent among victims with weakened immune systems, kidney failure, seizures, and coma.

Two species of *Ehrlichia* are currently recognized as causing disease in the United States: *E. chaffeensis,* first seen in 1986, and *E. ewingii,* the latest to be uncovered. They are carried primarily by the lone star tick (*Amblyomma americanum*), whose geographic range extends from eastern Texas through the south-central and eastern states. An illness initially called human granulocytic ehrlichiosis (HGE) and blamed on a third *Ehrlichia* species, *phagocytophila,* which was first described in 1994, has been renamed anaplasmosis (or HGA) since the pathogen was reclassified as *Anaplasma phagocytophila.* HGA occurs most often in the same states that have high incidences of Lyme disease because both are carried by the same black-legged ticks, *Ixodes scapularis* and *Ixodes pacifcus. E. ewingii* infections have been reported mostly in immunocompromised patients. A fourth species, *E. sennetsu* (reclassified as *Neorickettsia sennetsu*), that makes people sick with something called Sennetsu Fever was first identified in Japan in 1953 and is rarely found outside the Far East and Southeast Asia. It is probably caught by eating infected trematodes (flatworms) in raw fish.

• *Enterococcus* •

Several *Streptococcus*-like bacteria found in soil, water, and food, which commonly inhabit the human gastrointestinal tract, used to be called *Group D Streptococcus.* Since 1984, they have been classified in their own genus because they differ somewhat from proper strep germs. They can infect the heart, blood stream, urinary tract, and open wounds and are of special concern in various surgical operations. *Enterococcus faecium* and *E. faecalis* are especially worrisome, owing to their emerging super-resistance to antibiotics. They first appeared in New York City hospitals during the late 1980s and were held directly responsible for nineteen deaths out of a hundred infections (twenty-three other fatalities in the group had additional causes). Resistant strains soon turned up worldwide. By 1994, surveys indicated that 8 percent of all reported *Enterococcus*-caused illness was due to drug-resistant bacteria. The great fear among pathologists is that enterococci will transfer their resistance to other intestinal bugs, such as streptococci and staphylococci. The human gut is a perfect kitchen for this, making the dangerous outcome pretty much inevitable.

• enteroviruses •

The enteroviruses are a large mafia of viruses that inhabit the grand palazzi along the alimentary canal. The most infamous mobster causes polio. Others can trigger meningitis, encephalitis, respiratory illness, diarrhea, a severe form of conjunctivitis, and a dreary list of other afflictions, including the common cold. But all are known to produce inapparent (symptom free) infections far more often than active symptoms of disease. They are deceptive creatures, benignly tempered in most hosts but capable of vicious tantrums. No vaccine is currently available for the non-polio enteroviruses.

Besides the polioviruses, the group includes the Coxsackieviruses, named after the New York town where they were discovered during a 1948 polio epidemic; and the echoviruses, from the phrase "*enteric cytopathogenic human orphan viruses*," which were thought of as orphans when first identified during the 1950s in the sense that their relationship to human disease was unknown. "Enteric" means isolated from the intestinal tract. A certain enterovirus can produce many different symptoms that may be similar to those of another enterovirus or even of a virus from a totally different group. Researchers have tried to cut through this natural fog with so-called Virus Watch Programs, where families deemed normal from a medical perspective are observed for years and regularly tested in an attempt to match the state of their health with the presence or absence of specific germs. In addition, the enteroviruses are found so easily in sewage that much has been learned about how they infect various communities through testing of drainage systems (sewage treatment plants do not totally destroy viruses). Studies of this kind led to the startling knowledge that the ratio of inapparent to clinical polio infections in New York City was more than a hundred to one in the 1940s.

Enteroviruses are tough little nuts. They are unfazed by antibiotics and most disinfectants. They can be frozen for years and come out fighting. They can stand being heated to 122°F. They are found everywhere at all times in the tropics and subtropical regions but seem to prefer late summer or early fall in temperate latitudes. Chlorination will inactivate them, but the slightest presence of organic matter protects them, thus making water purification impossible in most settings. They crave human tissue—by the age of two, most children have had several harmless or mild bouts with enteroviruses. But fortunately for science they can be satiated sometimes with mice or monkeys. Highly communicable, only their habit of silent infection keeps them from being monsters capable of mass destruction.

The enteroviruses provide ample evidence that if an ecological niche opens up in nature, evolution will fill it. In this case, the usual expectation that poor children living in unsanitary conditions will suffer most is stood on its head. The frequency of severe outcomes of enterovirus infection, such as paralytic poliomyelitis, increases with age. The germs are usually spread by contaminated feces, so disadvantaged kids who lack good hygiene tend to get their first, immunizing infection early in life when they will probably experience minor malaise or no illness at all. As you climb the social ladder of cleanliness, first infection tends to be delayed to later childhood or young adulthood and beyond, when there is a greater risk for paralytic disease, meningitis, and other serious problems. Franklin D. Roosevelt was a robust thirty-nine-year-old aristocrat when he was permanently paralyzed from the hips down in August 1921.

Outbreaks of hepatitis A associated with eating bad shellfish can be due to enteroviruses dumped into bays and estuaries with sewage. Clams, oysters, and mussels collect the viruses as they filter food out of the polluted water. Even if water is not dirty enough to violate health regulations for bacteria content, it may contain viruses unperturbed by common treatment methods. Enteroviruses have been found along beaches miles from the nearest outflow pipe or offshore dump site. Since their favorite door into the human body is the mouth, swimmers are obviously at risk of infection. Testing for hazardous conditions is totally inadequate, making it wise for beachgoers to keep track of local construction projects.

A strong association exists between Coxsackievirus infection and insulin-dependent (type 1) diabetes, based on thirty years of epidemiological studies. Coxsackie outbreaks tend to be followed by an increased incidence of diabetes, perhaps because as the immune system fights off this virus, it inadvertently destroys insulin-secreting cells in the pancreas.

Enteroviruses are responsible for "hand, foot, and mouth" disease (HFMD) in humans, not to be confused with "foot-and-mouth" disease (FMD) in cattle. HFMD is a common, usually mild, childhood illness characterized by fever, mouth sores, and rash. Coxsackievirus A16 is the primary cause, but enterovirus 71 (EV71) has been responsible for outbreaks in Asia, in particular a major epidemic in China in 2008 that was especially dangerous for infants and young children. FMD is caused by a completely different group of viruses that are not a threat to people. (The reverse is also true: animals do not get the human disease.) Highly contagious among cloven-hoofed animals, including cattle, swine, sheep, and deer, FMD requires strict quarantine measures. The United States has been mercifully free of it since 1929.

• Epstein-Barr virus •

"Our spirits rushed together at the touching of the lips," wrote Tennyson in 1842, not realizing that far more tangible stuff might be part of the flow. Unpoetic physicians call this "intimate oral contact," a pleasure more widely known as kissing. In the 1950s, a decade renowned for adult attempts to suppress intimate contact of any kind among those who desired it most—that is, teenagers—a researcher named R. J. Hoagland observed that mononucleosis was spread by none other than intimate oral contact. Thus, though the medical term "infectious mononucleosis" had been used since the 1920s, the powerfully erotic phrase "kissing disease" entered our vocabulary, disseminated by *Reader's Digest* and *Time* far more thoroughly than science alone could ever accomplish. Mono was the sex scare of the fifties—tender evidence, perhaps, of how relatively innocent those days were.

The Epstein-Barr virus, named for the two scientists who discovered it in 1964, causes mononucleosis. It belongs to the primordial herpes family of viruses, whose members are infamous for their ability to persist or dwell latently for many years in their natural host, becoming active when the host's defenses weaken for any of myriad reasons. Teenage cavemen no doubt got mono, though it was one of the least of their health concerns. It is most common in the ribald fifteen- to twenty-five-year-old age group, probably because of more intense exposure, as the academics might say. In poor countries where hygiene is low, most children are infected by the time they turn six, but at this age the disease is mild or unnoticeable. Among the world's privileged classes, where saliva is abhorred, infection may more often cause serious problems—such as fever, jaundice, pharyngitis, or swelling of the spleen—because exposure tends to happen later in life. When infection occurs during adolescence or young adulthood, it causes mono about one-third to one-half of the time. Thus were Yalies linked with the hazards of make-out parties in the late 1950s and early '60s, when researchers at the medical school there found that three-quarters of entering freshmen were at high risk for mono because they arrived without antibodies. Connecticut is still the only state that requires reporting of infectious mono cases—a tribute, perhaps, to the enduring vigilance of the Westport way of life. In general, however, no one yet knows why some young adults who carry the Epstein-Barr virus get mono and others don't, which is perhaps just as well.

In the mid-1980s, a mysterious long-term malady characterized by severe fatigue, neurological problems, and other symptoms sometimes similar to mono became widely diagnosed under the rather dismissive term "chronic fatigue syndrome" (also called the derogatory term "Yuppie disease" because it was described as an upper-middle-class phenomenon). Some seventeen million people worldwide have the "syndrome," including at least a million Americans. Early studies discovered above-average levels of Epstein-Barr antibodies in sufferers' blood, but no exclusive causal relationship was shown.

Trying to control the virus—which infects as many as 95 percent of adults in the United States by age forty—is unrealistic, given the nature of intimate oral contact, and probably a bad idea, since it's better to be exposed during childhood than later. Vaccination (of high school students, say) would also be controversial, since there is a strong link between Epstein-Barr and cancer of the nose and throat in southern China, as well as a cancer known as Burkitt's lymphoma in Africa and Asia, a connection that is not yet fully understood but might be activated by vaccines. A statistical association has also been found between Epstein-Barr infection and multiple sclerosis, a disease rarely developed by people who do not carry antibodies to the virus.

• Fort Bragg •
(Leptospira)

In the summer of 1942, soldiers unlucky enough to be training at the Fort Bragg army base in North Carolina instead of peeping through beach house windows on Cape Cod came down with high fevers, headaches, rashes, and a bevy of other afflictions. They had picked up bacteria of the genus *Leptospira,* meaning "thin coil," which were later named after their beloved home away from home. After the war, investigators traced the disease, leptospirosis, to exposure by swimming in freshwater ponds and streams contaminated by livestock urine.

Long before this outbreak the disease had been noticed

among sewer workers, miners, farmers, and fish handlers. It was common in Japan, with mortality rates between 30 and 40 percent among thousands of victims every year. During the trench warfare of World War I, it plagued soldiers in Flanders. Rat piss was believed to be the culprit, but it is now known that myriad wild and domestic creatures—mice, dogs, bats, deer, foxes, raccoons, rabbits, pigs, goats, birds, frogs, snakes, fish, and on and on—play host to *Leptospira* and excrete zillions of microbes into the environment. The germs enter humans via direct skin contact. They are ubiquitous, making leptospirosis the world's most widespread zoonotic (animal-borne) disease. One to two hundred cases occur in the United States each year, mostly in the Southeast and Hawaii, where water temperatures are optimal for the bacteria's reproduction. But the largest recorded outbreak in the continental United States struck 110 participants in a triathlon at Lake Springfield, Illinois, during the summer of 1998.

When leptospirosis is severe it is called Weil's syndrome, after the German physician Adolph Weil, who first described it as "infectious jaundice" in 1886; it causes liver and kidney problems, meningitis, and vomiting. Antibiotics can control it, but the ecology of *Leptospira* and the number of species that can be hosts make prevention difficult.

• *Giardia lamblia* •

A protozoan parasite called *Giardia lamblia* will attach itself to your intestines with a little sucker, multiply like crazy, and make you fart. Have the gods no dignity? The creatures are found everywhere, but especially where sanitation is bad, since they move around in food and water polluted with human or animal feces, particularly those of beavers and dogs.

In 1674, the pioneer Dutch microscopist Antonie van Leeuwenhoek observed *Giardia* in his own feces without knowing it:

". . . animacules a-moving very prettily; some of 'em a bit bigger, others a bit less, than a blood-globule; but all of one and the same make. Their bodies were somewhat longer than broad, and their belly, which was flatlike, furnisht with sundry little paws, where-with they made such a stir in the clear medium and among the globules, that you might even fancy you saw a pissabed [louse] run-ning up against a wall; and albeit they made a quick motion with their paws, yet for all that they made but slow progress."

Scientific jargon has clearly demeaned the language since Leeuwenhoek's day.

Like so many other pests that take the fecal-oral route into us, *Giardia* (named for a French biologist, Alfred M. Giard) tends to infect kids in day-care centers, gay men, and tourists from rich countries who are oblivious to the realities of life in poor coun-tries. The United States is hardly free of the protozoans: *G. lamb-lia* (also known as *G. intestinalis*) has been found in 7 percent of stool samples tested nationwide, making it one of the most familiar intestinal infections. Fortunately, symptoms are usually mild or inapparent, but they can include weeks of bad cramps, diarrhea, and considerable loss of weight. Antibiotics can cure the illness over 90 percent of the time. The best way to avoid the parasite is by not drinking untreated surface water. If you have no choice, boil the water for at least a minute (three minutes at high altitudes) or pass it through a filter with an absolute pore size of one micron or less.

• hantaviruses •

Hantaviruses are a genus of viruses carried by mice, rats, and voles and transmitted to us by breathing air contaminated by their droppings, urine, or saliva. The first of these viruses to be isolated, in 1976, was Hantaan virus, the one that causes a bleeding disease called Korean hemorrhagic fever, which was first encountered during the Korean War; it was named after the Han River in South Korea. Hantaan had infected some twenty-five hundred U.S. troops, killing between 5 and 10 percent of them, and was feared as a potential biological warfare agent. The related Seoul virus causes a similar but less severe fever and, probably because it is carried by common rats, is found everywhere. Puulmala virus, carried by voles, is found most often in Scandinavia and western Europe. They all cause their victims' capillaries to leak blood, leading to the failure of crucial organs before the immune system can counterattack.

In 1993, the genus gained fresh notoriety when a newly recognized member was blamed for the deaths of thirty-two people out of fifty-three cases, primarily in the Four Corners region of the southwestern United States. Press attention focused on sudden fatalities among young, otherwise healthy Navajo Indians who suffocated after their lungs rapidly filled up with plasma from damaged pulmonary blood vessels. Within a month after the first death, the Epidemic Intelligence Service of the Centers for Disease Control and Prevention had identified a "new" hantavirus that they called the Muerta Canyon virus after a nearby geographic feature on the Navajo reservation—thereby offending the Navajos, who preferred to memorialize the "Valley of Death" as the scene of a Spanish massacre (the pathogen was renamed Sin Nombre, or no-name, virus). The American bug, which might have a fatality rate more than ten times higher than Hantaan, had been living for untold ages in deer mice (*Peromyscus maniculatus*, a ubiquitous North American field rodent), emerging from time to time to infect humans according to ecological whimsy. Traditional Navajo medicine men reported similar outbreaks during deer mouse infestations in 1918 and 1933. Indeed, Navajo legends recommended the burning of clothes soiled by mice.

Hantavirus pulmonary syndrome (HPS), as the disease is now known, struck 465 people in thirty states by the end of 2007. With a fatality rate of 35 percent, hantavirus would rank among the most lethal viruses ever found in the United States. Since the 1993 outbreak, several other hantaviruses have been found to cause HPS and named after the locales where they first appeared: Bayou virus in Louisiana (linked to the rice rat, *Oryzomys palustris*), Black Creek Canal virus in Florida (linked to the cotton rat, *Sigmodon hispidus*), and New York-1 virus (linked to the white-footed mouse, *Peromyscus leucopus*). Cases have been reported from Canada to South America, making HPS a pan-hemispheric disease.

There is no specific treatment or cure for hantavirus infection. Patients are simply rushed to intensive care units as soon as possible.

• *Helicobacter pylori* •

Helicobacter pylori is the only known organism that likes to live in human stomachs, an acid-rich barrier against so many dangerous germs. *Pylori* refers to the pyloric valve between the stomach and the duodenum, which was named after an ancient Greek word meaning "gatekeeper." Ignored for most of this century, *H. pylori* has been recognized only since 1993 as the prime cause of peptic ulcers, which afflict about 10 percent of Americans. Medical orthodoxy until then had deemed ulcers a noninfectious disease blamed on lifestyle factors such as spicy food or stress. To gather counterevidence, Barry Marshall, the physician in Australia who isolated the microbe in 1982, gulped a hefty dose of *H. pylori* and got an ulcer himself. For this revolutionary discovery and their courage in facing down the skepticism of medical orthodoxy, Marshall and his research partner, J. Robin Warren, won the Nobel Prize in Medicine and Physiology in 2005. *H. pylori* is more prevalent in older people and those living where sanitation is poor—more than three-quarters of the Third World population is infected, usually by the age of ten—

but there is no clear-cut reason why some individuals get stomach ulcers and others do not. What is new is the prescription of antibiotics for something that had long been believed to be unrelated to germs. *H. pylori* is also a potent carcinogen, leading to cancer of the stomach if the infection is untreated for twenty to thirty years.

• *Hemophilus* •

Hemophilus bacteria are normal, harmless passengers in our upper respiratory passages. One kind, however, is a prime cause of meningitis in kids under three: *H. influenzae* type b, or Hib.

At the turn of the century, researchers believed that *H. influenzae* caused flu; hence the name. But after flu was shown to be a viral disease in the 1930s, this bacterium was relegated to the role of secondary infection. When the germ's epidemic potential was recognized in the 1960s and '70s, vaccine development accelerated. Today, children are vaccinated beginning at two months of age, assuming they live in the relatively few places on earth where immunization is affordable and performed conscientiously.

• hepatitis •

The word "hepatitis" gets thrown around a lot in medical talk, because liver problems are so common. Many things can inflame the liver, often to the point of causing jaundice, the yellowing of the skin and tissues that is a telltale sign of liver disease, including alcohol, drugs, and other environmental chemicals of great category, as well as myriad microbes. Our modest purpose here will be served by zeroing in on the hepatitis viruses, which are unrelated but named after the first few letters of the alphabet: hepatitis A, B, C, D, and E.

Hippocrates was the first to note epidemics of jaundice, which certainly antedated his great acumen. John Dryden called jealousy "the jaundice of the soul" in 1687. The disease was often a factor in large military campaigns—during the Civil War, it struck at least seventy thousand soldiers. Until the middle of the twentieth century, hepatitis was pretty much synonymous with jaundice, which is actually just a warning that the liver can no longer cleanse the blood properly. By the beginning of World War II, investigators could differentiate between two types of hepatitis on the basis of how they spread—"infectious," via contaminated food or water, and "serum," via blood. It was not until the 1960s and '70s that the causative viruses were identified, partly with the help of a scandalous study of children at the Willowbrook State School for the Mentally Handicapped on Long Island, New York, where filthy living conditions had created a hepatitis infection rate near 100 percent.

Hepatitis A virus, the least dangerous, resembles the polio virus and consists of not much more than a bare strand of RNA genetic material in an icosahedral (twenty-sided) shell that reproduces only in the liver. It is known to infect only humans and a few other primates, mostly through contaminated feces. Hepatitis A is therefore a filth disease that tends to show up wherever sanitation and hygiene are poor. As with polio, people in poor countries tend to catch hepatitis A early in life, when symptoms are generally mild. Affluent travelers to such places may catch it as adults, however, and suffer badly. In the United States, where broad epidemics no longer occur, outbreaks are most common in day-care centers, among sexually active gay men, and in connection with polluted water or uncooked food, such as shellfish from contaminated water. Young infected children are almost never jaundiced but are prime sources of contagion for adults, who usually show all the classic symptoms of nausea, vomiting, dark urine, and yellowish eyes and skin. Since 1971, when hepatitis A hit some fifty thousand Americans, mostly in the fifteen- to twenty-four-year-old age group, the disease has trended downward in numbers and upward in age. Today it shows strong "socioeconomic" ties, a polite way of saying it strikes hardest at destitute minorities. In 2007, there were only 2979 cases reported in the United States of acute hepatitis A, but the annual number of new infections could be as high as 25,000. In May 2001, the Food and Drug Administration approved a vaccine called Twinrix against hepatitis A and B for persons over the age of eighteen. Two other vaccines, Havrix and Vaqta, that protect only against hepatitis A are licensed for children beginning at age one.

Hepatitis B virus is a much more complex bullet than A and naturally infects only humans, though it has close relatives that infect ducks, woodchucks, and squirrels. It can take as long as 150 days to incubate to the point of producing symptoms of disease, versus an average of 28 days for hepatitis A. It passes from person to person in blood, saliva, and semen, thus ranking it among venereal diseases. Before routine testing of the U.S. blood supply, thousands of deaths occurred every year from post-transfusion hepatitis of this kind. Moreover, the virus is extremely stable and can stay dangerous on needles, surgical tools, and even thorns or sharp stones. Hundreds of millions of people around the world are chronically infected and therefore capable of giving the virus to others. In Asia and Africa, hepatitis B is commonly passed from mother to child during birth or in breast milk. Because the germ's long-term presence in the body often brings on liver cancer, it ranks as the world's most common viral cause of cancer. An estimated 800,000 to 1.4 million Americans are chronically infected—the global total is perhaps 350 million—though the number of new infections reported annually has declined sharply from about 450,000 in the 1980s to 43,000 in 2007 (when 4519 acute cases were reported), probably because of routine vaccination of children since 1991 plus increased condom use and safer needle practices among gay men and intravenous drug users due to AIDS awareness. A vaccine was licensed in 1982 after being tested with the courageous help of New York City's gay community.

The hepatitis C virus, which until 1989 was so mysterious that it was referred to as "non-A, non-B," is the most common chronic bloodborne infection in the United States, encountered mainly in the context of intravenous drug abuse. It is related to the yellow fever virus and is a leading cause of chronic liver disease and cirrhosis. About 3.2 million Americans are chronically infected, with new cases (mostly due to shooting drugs) having fallen from an average of 240,000 a year in the 1980s to about 17,000 in 2007. There is no vaccine, though many biotech companies are trying to develop one.

Hepatitis D is a handicapped sibling of the others, so to speak. It can thrive only in cells also infected with hepatitis B, whose coating material it borrows while it boosts the severity of disease. If B were a jet warplane, D would be the afterburner. It is relatively uncommon in the United States. There is no vaccine.

Finally, hepatitis E virus, which travels from host to host via fecal-oral contact and contamination of water rather like hepatitis A, is recognized as causing major epidemics in Asia, Africa, and Mexico. It is especially dangerous for pregnant women. No vaccine is currently licensed, but an experimental vaccine under development since the mid-1990s shows promise in trials.

Happily, most cases of hepatitis go away by themselves with "favorable outcomes," as the docs say, though illness can drag on for a month or two. After such protracted misery, most of us would feel jaundiced about life indeed.

• herpes •

Herpes is one of the great virus crime families. There are more than ninety members that harass innumerable species, from us down to the lowliest fungi. They have been around for eons and are found in people everywhere, even in isolated primitive tribes. All have the capacity to survive for the entire lifetime of their host, hiding in nerve cells, sometimes producing symptoms of disease, often not. They have been implicated in causing cancer. Nothing can eradicate them from the body once they gain entry. Scary, yes, though generally not lethal.

The term "herpes," from a Greek word meaning "to creep," has been used for thousands of years. In popular speech it refers to just two of these thugs, *Herpes simplex* virus types 1 and 2, abbreviated HSV-1 and HSV-2. The first usually causes cold sores or fever blisters, first described by a Roman doctor, Herodotus, around A.D. 100 as "herpetic eruptions which appear about the mouth at the crisis of simple fevers." Recurrent episodes can be brought on by emotional stress, ultraviolet energy in sunlight, physical exertion, other infections, and immune system disorders. Shakespeare wrote of this unsightly curse in *Romeo and Juliet:* "O'er ladies' lips, who straight on kisses dream/Which oft the angry Mab with blisters plagues,/Because their breaths with sweetmeats tainted are." The second type causes similar spots in the genital area and was not publicly reported until sixteen hundred years later, by the French physician Jean Astruc (1684–1766). HSV-1 and HSV-2 are very closely related creatures; about half of their genetic codes are identical. Either type can infect the same body sites under certain circumstances. The fact that one is a social nonissue and the other a possible scandal is a perfect example of how our reaction to disease reflects cultural values.

Thus, studies have been concocted that found an HSV-2 infection rate of 70 to 95 percent among prostitutes and, lo and behold, zero among nuns. Surveys also indicate that two-thirds of people with either HSV-1 or HSV-2 show no apparent symptoms. Ergo . . . well, what exactly? A form of skin herpes named *Herpes gladiatorum* is found among college wrestlers and rugby players, who pick it up from mat burns and other repeated abrasions. So how about tarts who wrestle? Rugger-bugger nuns?

A drug called acyclovir was the first of a class of antiviral medications for controlling HSV outbreaks. These drugs are not cures, but they do reduce the number and duration of attacks in people who experience them. Debate about whether to make acyclovir available without a prescription has focused on the likelihood that resistant strains of HSV would spread more rapidly than otherwise. There is a great deal of demand: Though the percentage of Americans with genital herpes infections has decreased in the last decade, at least 45 million over the age of twelve—one out of five adolescents and adults—have suffered from them.

• **HIV** •

(human immunodeficiency virus)

Before descending into a welter of depressing detail, let it be said from the start that prevention is still the best way to fight the HIV monster, which has so far devoured some 25 million lives.

AIDS, the *a*cquired *i*mmuno*d*eficiency *s*yndrome ("syndrome" means the occurrence of a characteristic group or pattern of symptoms), has been widely called a modern Black Death. Would that this were so, because then a cure would be at hand. The analogy is valid, however, because of the state of fear and helplessness that AIDS has engendered since it was recognized in the early 1980s. AIDS is simply going to be around for a long, long time. After three decades of frenetic investigation at the forefront of virology, it is apparent that HIV and its effects on the immune system are not yet understood deeply enough to create a cure.

A disease as catastrophic as AIDS takes on myriad brands of sociopolitical baggage, clouding realities that are horrid enough on their own. The World Health Organization estimates that 33 million people were living with HIV in 2007, about 2 million of whom were children under the age of fifteen. Another 2.7 million became newly infected that year and 370,000 of those were under fifteen. Two million people died from AIDS. Although the epidemic is slowing down globally, with the incidence rate having peaked in the late 1990s, the number of infected people in crisis regions such as sub-Saharan and southern Africa continues to grow, with some countries staggering under Malthusean prevalences over 15 percent. More than 90 percent of AIDS cases occur in regions of the world where even mundane medical care is problematic. Since 1996, when a cocktail of several antiretroviral drugs became available to reduce the number of HIV particles in an infected person's body below the level that produces disease, deaths from AIDS have sharply decreased in rich nations, but less than a third of the people who need them in poor countries are getting them. In 2009, the Chinese government, which had dragged its feet in recognizing the disease during the epidemic's first decades, announced that HIV had become the nation's deadliest infection, as it is in many African countries.

In the United States, with more than 468,500 people living with AIDS, the disease is still overwhelmingly an affliction of men who have sex with men (53 percent of an estimated 56,300 new infections in 2006) and injecting drug users. But 31 percent of new infections in 2006 were blamed on high-risk heterosexual contact. Among racial and ethnic groups, African-Americans and Hispanics continue to account for the majority of reported cases (62 percent in 2006). The infection rate for black women was nearly fifteen times greater than that for white women and six times greater for black men than for white men. Through 2007, an estimated 583,298 people had died of AIDS, including 4,891 children under thirteen years of age.

These numbers are chronic dynamite because communal responses to epidemics are always easier to organize and sustain when the disease is perceived as being universal rather than marginal. Throughout history, diseases that disproportionately strike poor people (or people practicing what the larger society defines as vice) have been stigmatized to an extent that ultimately proves more damaging to society than the maladies themselves. Top federal officials in the United States originally opposed efforts

AIDS virus bursting from the membranes of infected T-4 lymphocytes.

AIDS virus particles found in a hemophilia patient.

to confront AIDS, apparently because of their bigotry toward gay men and drug addicts. Former Surgeon General C. Everett Koop's comments in 1991, reported in Lederberg et al., *Emerging Infections,* are often cited as testimony to this scandal: "Even though the Centers for Disease Control commissioned the first AIDS task force as early as June 1981, I, as Surgeon General, was not allowed to speak about AIDS publicly until the second Reagan term. Whenever I spoke on a health issue at a press conference or on a network morning TV show, the government public affairs people told the media in advance that I would not answer questions on AIDS, and I was not to be asked any questions on the subject. I have never understood why these peculiar restraints were placed on me. And although I have sought the explanation, I still don't know the answer." Despite the fact that AIDS research is now well institutionalized in the medical establishment, continued opposition to sex education, condom distribution in public schools, and clean-needle programs shows the reach of such "peculiar restraints," which can amount to a de facto death sentence.

Unlike the Black Death, which tended to come and go, AIDS may belong in the same realm as cancer or heart disease. That is, for some but not all populations it may settle in as an endemic killer. Or it may be similar to tuberculosis in the last century. It differs from all three points of comparison because it is totally preventable, communicable but not contagious. Of course, as a disease that is largely venereal, like syphilis, AIDS is subject to the debased mentality of divine retribution.

The literature on AIDS is vast and perhaps the fastest growing body of medical research. Interested lay readers can stay abreast with the journals *Science* and *Nature*. Briefly, HIV (*h*uman *i*mmunodeficiency *v*irus) is the third of at least six human retroviruses discovered since 1980 (there are undoubtedly more). Retroviruses contain an enzyme, reverse transcriptase, that converts viral RNA into a DNA copy that becomes part of the host cell's DNA. Before that, this class of pathogens had been connected only with various diseases in animals. HIV, which actually refers to two closely related viruses that cause AIDS in separate geographical regions, is part of a class of retroviruses known as lentiviruses traditionally associated with chronic arthritis and anemia. (Lentiviruses are retroviruses that cause slowly progressive, often fatal diseases.) HIV's unprecedented, tragic specialty is that it likes to attach itself to the very cells that turn on and amplify our immune system—the so-called T lymphocytes, specifically those with a protein called CD4 on their surface, which is the actual hookup point for HIV. Other lentiviruses do the same thing in cows and cats.

Once inside the T cell, the virus releases its genetic template (RNA) along with a chemical that allows it to be transcribed into the cell's own DNA. All offspring of the altered T cell thus contain the virus's genetic code. The T cell also may become a factory for new infectious HIVs, which destroy it as they burst out. The regulation of this life (or death) cycle is not fully understood, but one of the hallmarks of the process is the extremely long time—from two to fifteen years—that may pass between initial infection and onset of the syndrome (latency). Moreover, HIV constantly undergoes rapid genetic mutations in each individual victim, so that the immune system, or what's left of it, is always struggling to hit the right target.

The immune system fights a long, ferocious, but ultimately

losing battle against the AIDS virus. In late stages of the infection, victims lose and replace about 2 billion CD4 lymphocyte cells a day, while new virus particles appear at a rate between 100 million and 680 million a day. Other viral diseases, such as the flu or hepatitis, may also trigger warfare of such scope, but for a relatively brief time.

This is what makes the development of a single "magic bullet" vaccine so difficult. There is apparently no classic natural immunity (such as that seen after a bout with measles, say) that a vaccine could be developed to imitate, though some individuals somehow remain healthy for decades after infection. The virus is highly mutable, and a vaccine made for one strain may not protect against others. There is no satisfactory animal model for HIV infection that researchers can use before tests in humans. Clinical trials raise tortuous ethical issues. And they are enormously expensive—a six-year trial of a vaccine called RV 144 in Thailand involved 16,402 subjects and was hailed despite producing barely significant results (for unknown reasons), but it cost $105 million.

Where did this monster come from? The fact that there are similar lentiviruses in animals suggested a plausible source, leading researchers to a subspecies of chimpanzees native to west equatorial Africa: *Pan troglodytes troglodytes,* also known as the Central Chimpanzee. Bush-meat hunters were probably exposed to infected blood. (The earliest documented case of HIV infection was found in a 1959-vintage blood sample from a man in Kinshasa, Congo. Genetic analysis of this sample suggests that HIV might have evolved from a precursor virus in the late 1940s or early '50s.) Epidemiological evidence suggests that it traveled along the Mombasa-Kinshasa highway in Africa and thence to the United States during the mid-1970s, where it was first transmitted via anal intercourse among highly promiscuous (that is, hundreds of different partners a year) gay men in San Francisco and New York. By mid-1981 the Centers for Disease Control began to notice rare diseases—*Pneumocystis carinii* pneumonia and Kaposi's sarcoma—in young gay men that had previously occurred only in people with damaged immune systems. To other industrialized nations AIDS then became "the American disease." Because the virus passes easily in blood, intravenous drug users were the next vulnerable pool, along with the much smaller number of hemophiliacs and others who received contaminated transfusions. Since the virus can also go from person to person through vaginal intercourse (this has always been the primary route in Africa), the coterie of sexual partners of these initial victims was soon infected, too.

And so on and on. In the vicious AIDS circle, you essentially experience all of your lover's old flames. Pathogens adapted for this kind of infection tend to become permanent parts of the human condition, regardless of whether cures exist. Still, the authors of the most comprehensive survey ever of American sexual habits, released in 1994 under the auspices of the University of Chicago, concluded that HIV infection and AIDS are unlikely ever to become epidemic in the general U.S. population, because Americans tend to have sex with partners from similar social milieus. With few "bridges" between the marginal and the mainstream, the disease will not spread easily from those groups already hardest hit. Besides, more than 80 percent of American adults have only one sexual partner, or none at all, in a year's time, the survey found.

The risk of getting AIDS from casual, household, or workplace contact is still put at near zero, even from personal items such as toothbrushes, razors, clippers, and towels. Ditto for blood-carrying insects like ticks or mosquitoes. In addition to blood, the virus can be found in semen, breast milk, vaginal secretions, and saliva (though there is no evidence for catching it from an infected person's spit, as in coughing or sneezing). Outside the body, HIV is easily killed by common disinfectants like alcohol or peroxide.

From a sociopolitical perspective, there is no doubt that the horrendous AIDS death tolls of the 1980s and 1990s changed the tenor of life in gay capitals from San Francisco to New York to Provincetown. Regardless of one's sexual orientation, one had to recognize that AIDS moved the entire culture in conservative directions after decades of sexual revolution. But such is the force of great epidemics throughout history, for they rend the fabric of society in unpredictable ways.

Could the extreme mutability of HIV someday result in a contagious virus, like, say, polio or measles? While theoretically possible, the natural drift of evolution is toward lower virulence in a given population, not higher.

So let's end on that tiny note of optimism.

• influenza •
(flu)

Influenza, "flu" for short, is a universal human experience, familiar to most people as just a really long, bad cold. Flu is dangerous to the extent that it can lead to pneumonia, especially for the elderly, the malnourished, or individuals stressed by chronic lung or heart problems. The viruses that cause flu are wondrously prone to mutate, making the manufacture of vaccines an annual crap shoot. Usually nothing to worry too much about, flu is nonetheless potentially capable of enormous havoc. Even in a normal year, 5 to 20 percent of Americans catch flu, and about 36,000 die from it. When complicated by pneumonia, it is one of the top ten causes of death in the United States.

Influenza bugs are complex little machines by viral standards. Rugged spheres protected by layers of protein and fat, they contain long spiral chains of genetic material that look much like animal chromosomes. The spheres pirate a host by using more than 700 protein rods, called antigens, that fuse onto red blood cells and violate their membranes. In order to reproduce, the virus's chains must untangle, a process that offers many opportunities for "mistakes" to be made that might kill the offspring or make future generations more potent   or anything in between. The human immune system's antibodies are attracted to the rods, but the antigens may change or "drift" over time like a moving target. A successful bout against one variety of flu virus thus does not preclude a rematch with a slightly different opponent. If a new mutation of special strength pops up anywhere, resulting from a major change called an antigenic "shift," it will quickly spread around the world. All in all, an exasperating scenario for researchers, to say nothing of bedridden hosts.

Influenza epidemics are as old as human history, though hard to trace because of the generic nature of symptoms like fever, fatigue, coughing, and achiness. (Contrary to popular belief, "stomach flu" is a viral gastroenteritis not caused by the flu bug, though influenza can bring sick tummies.) The mysterious deadly Plague of Athens in 430 B.C. may have been partly due to highly virulent flu. Charlemagne's army may have been decimated by the disease during an epidemic in 876. There is unmistakable evidence of deadly influenza outbreaks during the sixteenth century, with the first recorded pandemic occurring in 1580 across Africa and Europe.

By 1647, influenza was reported in North America as an epidemic moved up from the Caribbean into New England, killing thousands. The disease was known by colloquial names ("la grippe," "jolly rant," "the new acquaintance") until after the epidemic of 1732–33, when an English doctor named John Huxham introduced an old Italian folk term that connected colds, coughs, and fevers to the astrological "influence" of the stars. Though it was a common killer generation after generation, flu did not attract serious medical attention during the centuries when other diseases were far more dramatic.

This situation changed with the great pandemic of "Spanish flu" in 1918–19, the twentieth century's worst of any kind and the deadliest faced by modern Western society. Spain was not fond of this term, needless to say, especially since the epidemic's geographic origin was puzzling and remains so today. Estimates of worldwide deaths start at 50 million out of perhaps 500 million clinical cases, amounting to one-third of the planet's population. India lost at least 12 million. The reliable figure for fatalities in the United States is 550,000. One-tenth of American workers were sick in bed during the winter of 1918. One-fifth of the U.S. Army and 28 percent of civilians caught the flu. More than 20,000 New Yorkers were killed in the fall of 1918. In Baltimore, all schools, churches, and theaters were closed; baseball games and Halloween events were banned; and the public was asked to hold off making phone calls because so many operators were ill. Wartime censorship of newspapers helped to curtail widespread panic. The virus was not satisfied with urban dwellers: Inuit villages in Alaska were wiped out, and Samoa lost 20 percent of its inhabitants. Physicians had little sense of why such a disaster was happening or what to tell people in order to avoid it. Katherine Anne Porter's short story "Pale Horse, Pale Rider" provides deep insight into the epidemic and the World War I society it ravaged.

Though an influenza virus had first been isolated from chickens in 1901, it was not recognized as such until 1955. There have been periodic waves of flu since 1919—"Asian flu" killed about 70,000 Americans in 1957–58 and "Hong Kong flu" about 34,000 in 1968–69—but we are probably still in an overall trough following the big one.

Not surprisingly, this disaster triggered sundry research in a no longer complacent scientific community. In 1931, an influenza virus was isolated from runny pig snouts, spawning the name "swine flu" and lending credence to the belief of some veterinarians that pigs had caught the flu from humans in 1918–19. The pig bug was, in fact, closely related to the human pest, but exactly what started the great epidemic may never be known, since no blood samples from the first wave of human infections were saved (precise viral identification has been achieved only for the second wave). So-called influenza A virus, now the most common, was found in the throats of infected people in 1933, influenza B in 1940, and the relatively mild and unusual influenza C in 1947. These are the three basic flu germs, variants of which are popularly designated according to where they first strike—Hong Kong (B), New Jersey (A), Bangkok (A), etc. Influenza A can change in both antigenic drifts and shifts, B only in drifts. Since 1918 influenza A has undergone two shifts, in 1957 and 1968. Nature's kitchen for new flu viruses is probably a duplex of pigs and waterfowl—yes, ducks and seagulls—that periodically pass them around to other creatures, including humans, in their guano. In 1980, an influenza A strain produced an epidemic in seals and caused conjunctivitis in people who tried to help them.

Since there is no reliable way to avoid influenza and no drug to make it go away once you catch it, vaccination is the only medical defense. Every year, an educated guess is made about what kind of flu will prevail during the next winter season. For example, an advisory committee of the Food and Drug Administration, after studying research on thousands of flu virus samples from 122 laboratories around the world, selected in February 2009 a vaccine against three strains expected to prevail during 2009–2010. They are named scientifically according to their type, origin, and protein structure: A/Brisbane/59/2007(H1N1)-like virus, A/Brisbane/10/2007 (H3N2)-like virus, and B/Brisbane 60/2008-like antigens. About 115 million doses were manufactured by six drug companies for those at risk of serious complications or anyone else who wished to minimize his or her chance of getting sick. Immunity to these strains develops within a week or two of getting the shot. (Because the flu shot is made from killed viruses, you cannot get flu from it. The nasal-spray vaccine called FluMist contains live viruses that have been weak-

*Gerald Ford's nemesis, the swine flu virus, properly called
A/New Jersey/8/76/ influenza.*

ened so as not to cause flu symptoms, but is approved only for
healthy people between ages two and forty-nine who are not
pregnant.) For people over sixty, flu vaccine can reduce the disease's incidence by half. Vaccinees who catch flu despite the shot
may experience milder symptoms.

In the United States, the first massive attempt to head off an
epidemic through general vaccination occurred in 1976 against
the so-called swine flu. President Gerald Ford told the American people that the 1918–19 virus was on its way back and
signed a law providing $135 million for an immunization cam-
paign that would ultimately reach only about a quarter of the
population. Along the way, the program attracted many prominent critics, including Ralph Nader and Albert Sabin, finally
collapsing when cases of Guillain-Barré syndrome, a rare paralysis, appeared at ten times the normal rate among vaccinees.
Happily, the flu epidemic never blossomed. Washington even-

tually paid some $93 million in damages, sending liability ripples through the courts that put a permanent crimp in the American pharmaceutical industry's enthusiasm for making *any* vaccines. In March and April of 2009, swine flu once again sparked a potent combination of scientific interest and public fear. Starting in Mexico, where thousands of people became infected and about a hundred died by the end of April, a novel strain of the H1N1 virus—the same basic influenza A-type bug that caused the deadly 1918 and weak 1976 episodes—spread into Southern California and quickly around the world. By June, the World Health Organization sounded its loudest alarm by declaring an international pandemic alert. By October, the virus had appeared in nearly 200 countries, infecting millions and officially killing almost 4000, with special risk to children and pregnant women. Fueling concern was the virus's knack for penetrating deeply into the lungs, thus causing more pneumonia than typical flu. The U.S. government spent about $2 billion to buy at least 250 million doses of new vaccine—the virus had been discovered too late to be included in the regular 2009–2010 combo—in order to protect more than half the citizenry. Along the way, a fresh strain of sociopathy arose among a rump element opposed to vaccination. Natural selection may eventually remove them from the herd, though they seem remarkably adept at passing their misconceptions from generation to generation.

Pigs are not the only animals sullied by human influenza names. So-called avian flu, from another influenza A virus found chiefly in wild birds and poultry, is caused by one of the few such bugs that can jump into humans: H5N1. Person-to-person spread is rare—the World Health Organization counted only 442 cases between 2003 and 2009, mostly in Asia, with 262 deaths (humans have little or no natural immunity, so infection can be highly pathogenic). The virus has not been detected in either birds or people in the United States. Vigilance is worthwhile, however, since subtle genetic variations could produce a more contagious organism. A vaccine has been available since 2007 and is stockpiled by the U.S. government.

By the way, flu viruses last for hours in dried mucus.

• *Legionella* •

Some diseases are said to have hastened the decline of empires, but only one is famous for putting a hotel out of business. In July 1976, an American Legion convention in Philadelphia's venerable Bellevue-Stratford Hotel set the stage for the discovery of a new genus of pathogenic bacteria. Later named *Legionella,* the bugs had found lodging inside 221 people who had come for the revelries or simply frequented the hotel. Thirty-four died subsequently from acute pneumonia. Understandably, the Bellevue-Stratford lost its charm and folded within a year.

Of course, *Legionella* was new only in the sense that humans had not been scientifically aware of it before. After five months of desperate detective work led to its identification, researchers took out old blood samples from previous mystery cases and verified that the same organism had caused similar outbreaks in a Washington, D.C., hospital in 1965 and a Pontiac, Michigan, health department office in 1968 (thereby proving, among other things, that germs are not afraid of medical facilities). They even thawed out a guinea pig from 1947 that had been injected with blood from a patient who died from something untraceable at the time and found *Legionella.* "Legionnaire's disease" has probably been behind thousands of deaths from pneumonia every year in the United States for a long time, especially since air conditioning became part of urban life and perhaps since the unwashed masses accepted indoor plumbing. Between 8,000 and 18,000 Americans get it annually, with a fatality rate ranging between 5 and 30 percent.

The culprit in Philadelphia was specifically *L. pneumophila,* one of some forty-three *Legionella* species now known to exist around the world, at least nineteen of which cause pneumonia in humans. *L. pneumophila* and its brethren like to live in environments akin to puddle scum—wet, dark, and kind of rotty. At the decidedly un-rotty Bellevue-Stratford, air-conditioner cooling towers on the roof provided a conducive summertime abode, from which the germs circulated throughout the edifice in a fine infectious mist. But *Legionella* bacteria have also been found in showerheads and faucet pipes. *L. pneumophila* can survive in room temperature tap water for more than a year. Streams, ponds, and plain old mud may harbor them. Hot tubs, humidifiers, vegetable stand misters, decorative fountains—the list of possible habitats is unsettling. Chlorination may not remove the threat if the concentration of organisms is high.

Once a person becomes infected by inhaling enough bacteria, two courses of disease (properly known as legionellosis) may occur, depending on such factors as the victim's age, use of cigarettes or alcohol, and other conditions that are still unclear. Legionnaire's disease, the more serious, strikes 5 percent or fewer of people exposed to *Legionella,* takes two to fourteen days to develop, and kills up to 40 percent of its victims. The comparatively mild Pontiac fever shows many of the same symptoms (high fever, chills, headache, cough) but does not include pneumonia, hits 90 percent of those exposed, comes on in just one to two days, and may be debilitating for a week or so but is not fatal.

Neither disease spreads person to person—you need to have breathed the original mist of contaminated water, and quite a bit of it. Thanks to the unlucky Bellevue-Stratford, regulations concerning the maintenance of big air-conditioner systems are more strict today than in 1976.

• *Listeria monocytogenes* •

The bacterium *Listeria monocytogenes* causes food poisoning that can be potentially fatal if it leads to meningitis. *Listeria* are found in soil and water everywhere and inhabit the gut of humans, birds, spiders, livestock, and other animals. The disease known as listeriosis is not generally of concern for otherwise healthy individuals but is dangerous for pregnant women, infants, and anyone with a weakened immune system. It is relatively rare in the United States, striking perhaps 2,500 people every year.

L. monocytogenes has the especially bothersome ability to thrive on refrigerated foods and therefore is sometimes encountered in cold delicatessen goodies. In October 1994, the U.S. government issued a warning against eating potato salad manufactured by I&K Distributors in Ohio and sold in twelve states, due to *L. monocytogenes* contamination. In 1998, an outbreak that struck forty people in ten states, causing three deaths and one miscarriage, was traced to bad hot dogs made by Bil Mar Foods.

Antibiotics can control it.

• mad cow •

Cows get mad cow; humans get Creutzfeldt-Jakob. The relationship between these two fatal brain diseases has been under intense scrutiny since 1995, when an unusually high number of human cases became apparent in the United Kingdom, where mad cow had been epidemic in cattle since the mid-1980s. By October 2009, British investigators had tallied 167 primary cases of this so-called new variant Creutzfeldt-Jakob disease (vCJD) with 162 deaths. Three more fatal cases in Britain were associated with blood transfusions. France had the next highest total of 25, with 24 deaths. Nine other countries were in the low single digits of this tragic toll, including the United States with three fatal cases all linked to exposure to BSE-infected cattle products while living in the UK (two cases) or Saudi Arabia.

In its classical form, Creutzfeldt-Jakob—or "transmissible spongiform encephalopathy" (TSE)—was first described in the 1920s by two German neurologists as an extremely rare, sporadic affliction with no identifiable cause, mostly among the elderly. The new type showed a similar, invariably lethal degeneration of the brain but focused on young people under thirty. Mad cow—or "bovine spongiform encephalopathy" (BSE)— perhaps originated from scrapie, a brain disease in sheep and goats that had been known for centuries, characterized by spongy holes in the brain that bring loss of coordination, paralysis, and death. Scrapie also began to rage through British cattle during the 1980s.

Until the practice was banned in 1988, cattle in the United Kingdom were fed with a meat and bone meal product made from the rendered carcasses of livestock, including sheep, as a nutritional supplement. The pathogen responsible for BSE apparently contaminated this feed, then the healthy cattle, then the humans who ate infected beef.

In fact, small studies from the 1970s had already found that TSE victims were more likely than other people to have dined upon animal brains. So what was the pathogen? First thought to be a "slow virus," the culprit is instead believed to be a new, subviral particle called a prion, a self-replicating protein that contains no genetic material. Prion proteins are normal components of cell surfaces, but pathogenic forms damage brain cells. The butchering process in slaughterhouses can spread infected brain tissue all over an animal's carcass, resulting in prion-infected meat on the dinner table. The 1997 Nobel Prize in Medicine was awarded to the discoverer of prions, neurologist Stanley B. Prusiner.

Through May 2009, only 20 cases of BSE have been identified in North American cattle: 17 in Canada (one imported from the United Kingdom) and three in the United States (one born in Canada). Even in the United Kingdom the risk of getting the new variant of Creutzfeldt-Jakob disease from eating beef is considered very low, happening perhaps once out of every ten billion servings. To reduce the risk even further, American travelers can choose to eat beef only in the form of solid pieces of muscle meat, such as steaks (not T-bone!) instead of burgers or sausages more likely to contain prion-infected tissue. And skip the *cerveau*.

• Marburg/Ebola viruses •

Marburg and Ebola are two exotic filoviruses—the only known members of this family, in fact—that can earn big bucks in Hollywood. They guarantee high fatality rates (up to 90 percent in large outbreaks), produce grotesque symptoms, lurk in darkest Africa under murky circumstances—in short, possess all the ingredients for a blockbuster. They have already triggered far more melodrama than disease, as a result of two famous incidents when imported laboratory monkeys carried the microbes into the rich white man's world.

One day in the summer of 1967, three employees at a vaccine factory in Marburg, Germany, complained of flu; by the next day their condition deteriorated into what became weeks of acute viral blood poisoning, bizarre rashes, severe diarrhea, bloody vomiting, and hemorrhages from every orifice. Also, their skin came off. In Germany and Yugoslavia (now Serbia) the number of cases among pharmaceutical workers, the doctors who treated them, technicians, and household contacts climbed to thirty-seven. Seven died by the end of the year. Others suffered liver damage, impotence, and psychosis. Investigators discovered that all the cases were linked by exposure to a shipment of African

Human liver infected with the Ebola virus.

green monkeys from Uganda, half of which had died en route. They found a large new rod-shaped virus in the monkeys' tissue that resembled a woolly bear caterpillar, one that could roll up tightly for protection just like that cuddly little creature. No one has ever figured out exactly where the virus comes from, how humans initially catch it, or much of anything else that you would like to know about such a monster. Methyl alcohol knocks it out.

In the summer of 1976, panic swept remote villages in Zaire (now the Democratic Republic of the Congo) and Sudan in the wake of sickness and death in a form similar to the Marburg cases, only worse. Local physicians at first suspected yellow fever or typhoid, then Lassa fever, but soon called for outside help. Investigators discovered a virus shaped like a question mark, surely one of nature's all-time worst jokes. Named Ebola after a river near the epidemic's starting point, the virus turned out to be closely related to the Marburg germ and just as mysterious. Cases mushroomed to 318 in Zaire, with an 88 percent fatality rate. In Sudan the virus species was perhaps somewhat weaker, hitting 284 people, with a 53 percent death rate. Both species have reappeared in humans from time to time in the Congo, Gabon, and Uganda. A new species called Ebola-Bundibugyo caused an outbreak in Uganda in December 2007, with 149 cases and 37 deaths.

In 1989, the Marburg plot recurred in the tony Washington suburb of Reston, Virginia, when monkeys imported from the Philippines starting dropping like flies at an animal quarantine facility. A new Ebola species was soon isolated, causing feverish prayers among lab workers that the nation's capital would be spared, until it became evident that this new strain was not a man-killer—several people were infected by the "Reston" virus, but none showed any symptoms. The scary episode further besmirched the reputation of imported monkeys and made a small fortune for at least one magazine writer. Four times since then, Ebola-Reston has been found in monkeys from the Philippines under similar circumstances, but it has caused no problems in humans.

A Centers for Disease Control worker exposed to the Ebola virus being loaded onto an airplane while isolated in a plastic tent.

In addition to underlining the necessity for strict international controls on animal trafficking and the potential for vicious diseases to escape from remote areas through fast modern transportation, the Marburg and Ebola outbreaks demonstrate one of the basic tenets of infectious disease: If a germ is exceedingly deadly, it will flash through a population rather quickly rather than cause massive, prolonged epidemics. Germs need fresh, healthy hosts, not corpses.

• measles •

Measles, or rubeola, is *the* most infectious common disease—one person with measles who blunders into a crowded room will give it to almost everyone. If there were no immunization, measles would be universal. Epidemics used to happen like clockwork every two to three years. That the United States, which declared endemic (native) measles eradicated in 2000, nonetheless continues to experience significant outbreaks that are mostly traceable to imported cases but sometimes blameable on unvaccinated American children—who are often victimized by anti-vaccination parents—is stark proof of the absolute necessity for maintaining high immunization rates.

Measles is caused by one of the paramyxovirus gang, others of which bring mumps and sundry respiratory illnesses. The virus travels from victim to victim primarily in fine droplets sprayed by coughing and sneezing. About two weeks after exposure, the well-known rash appears with high fever. New viruses will grow in cells throughout the body, but the mucous linings of the respiratory tract are their favored breeding ground, providing a perfect launchpad. Particularly in Africa, measles causes deaths from pneumonia, diarrhea, and malnutrition. People who recover are immune for the rest of their lives.

A thousand years ago, the disease was so ubiquitous among

children that its earliest reporter, the Persian physician Abu Bakr Muhammad ibn Zakariyā, known as al-Rhazes of Baghdad (850–923), believed it was simply a natural episode of childhood—like losing baby teeth—rather than an infectious illness. The measles virus is not carried by any animal other than humans, and it needs a local population of between 300,000 and 400,000 to set up a sustaining, continuous supply of virgin bodies. Otherwise it dies out. Such settlements did not exist before the Sumerian cities in the Middle East, around 2500 B.C. Thus, medical historians theorize that measles evolved sometime after this date from a virus in cattle or dogs. Neither Greek nor Roman writers made note of it, but it might have been classed together with other severe fever-rash afflictions. Al-Rhazes was the first to distinguish between measles—which he called *hasbah,* meaning "eruption"—and smallpox, though he thought they had the same cause, a view that persisted for nearly eight hundred years. The word "measles" did not appear in English until the fourteenth century, stemming from the word "miser," which was used to refer to the wretchedness of lepers.

The disastrous effect of this germ on previously unexposed populations was most spectacular during the colonial era. It was instrumental in cutting the population of central Mexico from about thirty million to three million within fifty years after Cortés's arrival. In otherwise healthy, well-nourished children, measles is not usually a killer. Yet eight hundred children died in the Charlestown neighborhood of Boston in 1772. In 1848, measles struck Hawaii for the first time, apparently imported from California, killing 27 percent of some 150,000 islanders who came down with the disease. If these figures sound incredible, one need only consider the arrival of the H.M.S. *Dido* in the Fiji Islands in 1873. Within three months, a quarter of Fijians, some 30,000 people, were dead from measles. The virus is known for remarkably consistent potency: In 1951, measles commenced with a single case among 4,300 never-before-infected

A Japanese woodcut, circa 1900, showing measles as an especially hirsute demon.

Greenlanders, and within six weeks all but 5 of them caught it (the mortality rate was only 1.8 percent, thanks to proper care).

Measles vaccine was not approved in the United States until 1963, when a half-million cases were expected annually. No doubt because of the disease's close historical association with smallpox, a technique had been attempted as far back as 1758 that was similar to Edward Jenner's inoculation of healthy individuals with pus: Blood from measles patients was smeared on scratched arm skin. This crude measure was not as successful as equally primitive smallpox prevention with pus, however. Researchers were unable even to grow the virus outside the human body until the late 1930s. By the mid-1950s, they had learned how to measure how much virus was present when cultivated in chick embryos, thus opening the way for production of a safe vaccine.

Unlike the smallpox virus, which has been eradicated in nature, the far more infectious measles virus still lurks everywhere ready to strike the nonimmune. In 1977, the United States set a goal of conquering measles by 1982. Reported cases fell to fewer than 3,000 a year, but then shot up to 25,000 in 1990. In the spring of 1994, an infected skier in Colorado passed the disease to hundreds of unvaccinated students, many of whom were Christian Scientists, who shun medicine in favor of prayer. By 1999, reported cases had declined to 100, 66 of which were either contracted outside the country or due to exposure to those imported cases.

All children should be vaccinated at twelve to fifteen months (a bit younger in developing countries) and get booster shots later, but inadequate government support for public health has made this unachievable in many countries. Because immunization programs must be nearly perfect to counter the virus's extremely efficient reproduction, poverty alone can easily derail even the best intentions in developed countries. About 20 million measles cases occur globally every year, causing some 242,000 deaths among children.

A disease similar to measles, but comparatively benign and less infectious, is caused by the unrelated rubella virus. Also known as "German measles" because of research done by German scientists, it would be of little consequence if not for the fact that unborn babies are at high risk of birth defects (at least 20 to 25 percent) if their mothers contract the disease during the first trimester. This was discovered only in 1941 by an Australian ophthalmologist, Norman Gregg, who reported a link between mothers who had caught rubella and infants with congenital cataracts. After centuries of considering rubella to be trivial, the medical establishment balked for years at Gregg's hypothesis, even attacking him editorially in the prestigious medical journal *The Lancet* in 1944. But further research proved that rubella could indeed cause a bevy of defects, including heart problems, deafness, language disorders, and bone lesions.

Rubella virus.

Large epidemics continued to occur regularly around the world until a vaccine became available. Production of a vaccine was first licensed in the United States in 1969, when about 50,000 cases were occurring every year, and has since reduced the disease to negligible importance. As long as kids get their MMR (measles, mumps, rubella) shots, the reservoir of germs that might infect women of childbearing age will remain nearly dry.

• mumps •

Mumps is the disease with a name that sounds just like the way it makes you feel. Hippocrates described an epidemic of mumps, well-known across the ages for painful swelling of the parotid (salivary) glands in children and, in about a fifth of older male patients, of the testicles. Like measles, mumps is caused by a paramyxovirus transported into the respiratory tract via fine droplets from an infected person, but it is far less dangerous and contagious than its cousin, being similar to flu or rubella.

Mumps virus.

Mumps comes and goes with no apparent symptoms in about a third of all cases. Contrary to popular fear, it rarely causes sterility in adult men, though many of them unlucky enough to catch it must swear that's going to happen. It hurts. Because the virus likes glandular and nerve tissue, mumps can be accompanied by complications in many organs—pancreas, brain, heart, thyroid—but they are seldom serious.

Before a vaccine became available in the 1960s, mumps tended to be an urban affliction with fairly irregular occurrence. It was thus bothersome to the military whenever large numbers of men were mustered together from different locales—many had already been exposed and were therefore immune, but many were susceptible. While not capable of causing significant fatalities, mumps could be a dangerous nuisance for an army in the field. During World War I, some 231,490 cases in the U.S.

HARPER'S WEEKLY.

A MEMORY OF THE MUMPS.

Army resulted in 3,884,147 man-days of lost duty. Mumps ranked third, behind flu and gonorrhea, for putting otherwise healthy soldiers out of action. There were even secondary attacks, especially among buglers. Discovery of the mumps virus in 1934 brought better understanding of how it was transmitted. The Army had only 103,055 cases during World War II in a vastly larger force, probably because the general population of recruits was less rural and therefore had been more evenly exposed before being drafted.

Mumps vaccine is usually given to one-year-olds in combination with measles and rubella vaccines (MMR). Reported cases in the United States declined during the 1990s from about 6,000 a year to 350, probably because of a recommended second dose of MMR before kindergarten or first grade. A worldwide immunization campaign could relegate mumps to archival status.

• mycobacteria •

Two organisms of the genus *Mycobacterium* are infamous for causing tuberculosis and leprosy, but there are myriad other types. They live harmlessly in soil, water, insects, and synergistically around plant roots, and in many animals besides humans. Infections in people are probably due to contact with dirt, dust, or water. In children, old people, and particularly the immune-impaired, they sometimes cause lymph-system problems, skin ulcers (following cuts received in pools, fish tanks, or similar places), and, rarely, bone, urinary tract, and nerve disorders. These last, more serious infections are AIDS alarms attributed to the so-called *Mycobacterium avium* complex (or MAC, composed of *M. avium* and *M. intracellulare*).

• *Neisseria gonorrhoeae* •
(gonorrhea)

To the ancient Chinese, it was the disease that is "different from all others." To the Greeks, it was *strangouria,* or urinary blockage. The Hebrews, who went on poetic record as being extremely wary of loose women and adventuresses (Proverbs 5), referred to it as "discharge" (that's the polite translation, at least). The French, ever graphic, called it *chaude pisse*—hot piss—as far back as the twelfth century. The worldly Dutch said *klapoor,* from which the English got "clap." It is gonorrhea (from the Greek words *gonos,* meaning seed, and *rheein,* to flow), one of the oldest and most familiar diseases of humankind. "Pissing razor blades" sums it up quite well.

In the dark spectrum of venereal (from Venus, the goddess of love) diseases, gonorrhea has a reputation for being not so bad. It can ruin a relationship, but it won't really hurt anybody. This

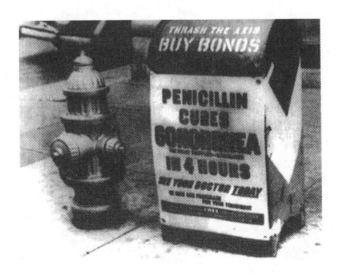

misunderstanding is probably a male view, ascribable to the fact that women suffer gonorrhea's worst consequences. The picture is clouded, moreover, by three common suppositions: Most infected men experience mild discomfort and then get better automatically; antibiotics are available when necessary; and most infected women are symptomless. All are true, but there are important qualifications.

More than 90 percent of the men and women who infect each other will have no symptoms at the time. Mathematically speaking, the odds that infected males will transmit the disease to their sexual partners are between 50 and 70 percent. Some 20 to 40 percent of women, once infected, will develop pelvic inflammatory disease—a no-nonsense complication that can lead to infertility—or develop what physicians ominously call a

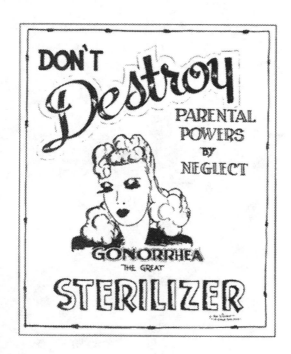

"constellation" of other gynecological problems. Women are less likely to give the disease to their uninfected partners (the odds are 20 to 30 percent, still way better than any casino game). But some 80 percent of men, once they've caught the bug, will then get what the ancient sages warned about. (Lesbians rarely give gonorrhea to each other. Gay men do, and may suffer from infections of the rectum and throat.)

It should be added here that nonsexual transmission of gonorrhea is so unusual as to be freakish. The bacterium that causes this illness, *Neisseria gonorrhoeae* (named for Albert Neisser, the German bacteriologist who discovered it in 1879), is exquisitely well adapted to certain human tissue and ways of finding fresh supplies, hence its primordial parasitic relationship. Infants can pick it up in their eyes during birth from infected mothers, and it used to be a prominent cause of blindness in newborns, but routine application of silver nitrate solution to the eyes of every child at birth has largely overcome this tragedy. Prepubescent children with gonorrhea have almost certainly been sexually abused.

Until 1937, when sulfa drugs were found to be effective killers of *N. gonorrhoeae,* there was no cure. Penicillin soon joined the fray. Unfortunately, the pest is very quick at learning how to resist antibiotics. Sulfa drugs were used so freely during World War II that resistant microbes popped up everywhere. By 1976, after the mushrooming of venereal disease that accompanied the advent of the Pill and the sexual revolution, penicillin-resistant strains (known as PPNG, for *p*enicillinase-*p*roducing *N*eisseria *g*onorrhoeae) began to appear in the United States, imported largely from the Philippines by military personnel. Microbes soon popped up that were also resistant to tetracycline, spectinomycin, and fluoroquinolones. Other drugs can still combat *N. gonorrhoeae,* but the trend is obvious. There is no natural immunity, nor any vaccine. In 2007 some 356,000 cases were reported in the United States—mostly among young

African-American men and women—at a rate that has been stable for about a decade, but this number is probably just a fraction of the annual incidence.

Yet another excellent reason to use condoms.

• *Neisseria meningitidis* •

In 1974–75, a vast epidemic of meningitis, a disease in which membranes surrounding the brain and spinal cord become painfully inflamed, blanketed Brazil and was especially cruel to its poorest citizens in the squalid barrios of São Paulo, Rio de Janeiro, and other cities. Many victims fell desperately ill in minutes, with stiff neck and fever leading to hemorrhages, coma, and death within a day. The military junta responded with a vaccination campaign that reached some 80 million people—an unparalleled feat perhaps possible in such an unwieldy society only when ordered by generals—but not before *Neisseria meningitidis* bacteria had sickened more than 250,000, killed more than 11,000, and left a third of the survivors neurologically damaged.

Such paroxysmal epidemics of *N. meningitidis* are uncommon. The disease is not as contagious as colds or the flu, but because the microbes spread easily in overcrowded living conditions, they tend to claim lots of victims once they get started. An epidemic in Nigeria in 1948–50 took 14,273 lives out of 92,964 reported cases. The attack rate in Chad reached an incredible 11,000 per 100,000 people. In children who were not treated with antibiotics, mortality rates sometimes topped 80 percent. Much like the polio virus, *N. meningitidis* produces far more "silent" carriers than symptomatic infections, a factor that helps fuel hysteria in populations already shaken by the precipitous, seemingly random appearance of gruesome cases. The bacteria have also caused extensive out-breaks during modern military mobilizations, as experienced by American recruits during both world wars, Korea, and Vietnam. The disease was of such concern to the British Army during World War I that studies were done to determine the precise distance that bunks should be separated to minimize conta-gion. But like all types of meningitis, this one always focuses on children.

Though epidemics probably occurred at roughly twenty-year intervals in the nineteenth century after meningococcal disease was first described in Switzerland in 1805, various types of *N. meningitidis* were not identified until the 1940s and '50s. Wide-spread use of antibiotics followed these studies, but resistant strains appeared by the 1960s. The Brazilian epidemic was com-plicated by the fact that two types of *N. meningitidis,* known as A and C, rampaged simultaneously—type A was new to Brazil and treatable with antibiotics, while type C has been around before and was now highly drug-resistant. In the modern age of jet travel, this scenario could recur in any big city with dense, impoverished neighborhoods. Hundreds of thousands of cases have occurred in the so-called meningitis belt of West African countries in recent years.

Like the vaccine used in Brazil, which came from France, the current vaccine in the United States—MCV4, licensed in 2005 and recommended for routine vaccination of everyone between the ages of 11 and 18 at the earliest opportunity, as well as older people at increased risk, such as military recruits and college freshmen living in dorms—is effective against A and C bacteria, plus two other types designated Y and W-135. Because A bacte-ria are now rare in the U.S. and there is no licensed vaccine against B, some one to two thousand cases of invasive meningo-coccal infection still occur every year, with a death rate of 10 to 14 percent despite treatment with antibiotics.

• papillomaviruses •

"Barley-corn, barley-corn, Injun meal shorts,/Spunk water, spunk water, swaller these warts." Thus did Huckleberry Finn try to do away with the most familiar result of papillomavirus infection, the common skin wart.

Comparatively unpoetic scientists had described the viruses and associated diseases by the middle of the twentieth century, but because of difficulties in culturing them in the lab, research did not blossom until the 1980s. Over a hundred types of papillomaviruses are now known—most of them harmless—about thirty of which are spread by sexual contact. An important link has been discovered between infection with certain ones and cervical cancer, which causes more than four thousand deaths every year in the United States. Moreover, evidence is accumulating that sexually transmitted types of papillomavirus somehow interact with *Herpes simplex* virus type 2 and HIV to increase the risk of cervical cancer. The picture is not clear, however, since 10 to 50 percent of healthy women carry papillomavirus.

This virus has an affinity for skin and mucous membranes, hence the Huckleberry Finn bumps and the less well-known venereal warts that can appear on the genitalia and anus. They are transmitted by direct contact and take an average of two to three months to produce a visible wart. Skin warts can be easily removed by freezing them with liquid nitrogen, but other kinds require more finesse, shall we say. So, yes, use a condom.

• parainfluenza viruses/respiratory syncytial virus (RSV) •

As any parent knows, a group of little kids at the playground will usually include gigglers, criers, screamers, and coughers.

Fortunately, we are concerned here with coughing, because it is far easier to explain than the rest of the din.

Bronchitis, bronchiolitis, croup, and pneumonia—four main stays of the pediatrician's trade—are brought primarily to young children by members of the Paramyxoviridae family of viruses, others of which cause mumps and measles. These diseases are universal because the bugs are, too. They are highly infectious through personal contact and need invade our bodies no deeper than our noses or throats to start replicating like crazy in the mucus there. The only way to avoid them would be to stop breathing. They strike early in life because they are *out there.* Most people grow partially immune to them, which is why parents at the playground aren't hacking as much as the kids.

The parainfluenza viruses—there are four types of medical interest—cause lower respiratory diseases in kids and upper respiratory problems in adults (most grown-ups have some immunity, so the bugs are defeated before they can travel very deep). They were first isolated in the mid-1950s from children with croup. The two types most closely associated with croup tend to strike tots under two in the autumn of the year, with the greatest number of cases in populous areas coming at two-year intervals. Boys seem to get croup more often than girls. The type associated with bronchiolitis and pneumonia works all year round, mostly on babies under six months old, with no preference for blue or pink. So far, attempts to develop vaccines, perhaps best administered as a nasal aerosol, have not met with much success. Happily, they're easily killed with soap and water, so handwashing is a good defense.

The respiratory syncytial virus (RSV) was first found in a chimp with a cold. It is now recognized as the prime agent of life-threatening diseases of the lower respiratory tract (pneumonia, bronchiolitis) in infants and children all over the world. It is so universal that all adults are believed to carry an antibody, as do babies for the first few weeks of life until the resistance acquired from mother wears off. RSV is the only virus that causes pediatric epidemics of respiratory disease every winter in big cities. As with so many other ills, poor kids are more likely to suffer from serious infections, perhaps because affluence somehow helps delay exposure past the period of greatest vulnerability (the first six months of life) to an age when reactions are milder. High-priority efforts to make a vaccine have been about as futile as those against the parainfluenza virus. The matter is complicated by the fact that an infant's defensive system against RSV invaders is suppressed by RSV antibodies carried out of the womb, so that any vaccine would have to finesse this obstacle at least until those antibodies decayed. Handwashing helps, Mom and Dad.

• *Plasmodium* •
(malaria)

Worldwide, about 300–500 million people suffer every year from malaria, and more than a million die. The disease is caused by four species of this little parasitic animal. We can venture to use the word "animal" here, since *Plasmodium* is a genus of protozoans, considered to be higher on the evolutionary ladder than viruses and bacteria. At one stage in plasmodia's curious reproductive cycle they even come in two sexes. Malaria—from the Italian *mal aria,* meaning "bad air"—is of primary concern throughout tropical Africa, where more than 80 percent of the world's cases and 95 percent of deaths occur, and in rural areas of Asia and Latin America. But it is not entirely absent elsewhere, including the United States, where there are about 1,500 cases every year, usually among immigrants and international travelers. The sheer number of infections and fatalities, mostly among children, represents a staggering toll for a disease most of us probably think of as belonging in the distant past, one that

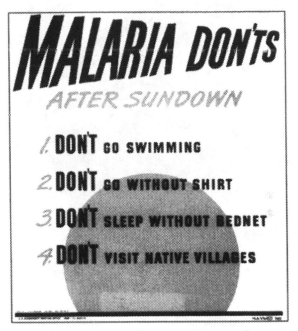

And don't ever bother to come here again.

even experts believed was headed for extinction not long ago. In many ways, malaria is stubborn.

Clear references to an affliction of intermittent fevers can be found at the start of recorded history in ancient Chinese and Sanskrit. Hippocrates described it in clinical detail; historians think it played a significant role in the cultural decline of Greece. The disease was so ferocious around Rome and along the west coast of Italy that some areas remained uninhabitable until the twentieth century. In Africa, it prevented European explorers from making contact with indigenous peoples as surely as any uncharted jungle or desert. More than 40 percent of British soldiers in India during the mid-nineteenth century suffered from malaria.

Since the 1890s, malaria has been understood to be spread not person to person but via mosquitoes, specifically certain

Red blood cells infected with Plasmodium, *an agent of malaria,*
magnified a thousand times.

female *Anopheles* species which feed on warm-blooded animals.
The males are innocent vegetarians. (For ages, people were well
aware that living around odoriferous swamps and low areas
could bring on deadly "black-water fevers" or "agues," an
empirical observation that was part of the "miasmatic" theory
of disease.) While these spindly little Nimrods—who do a dis-
tinctive headstand with their hind legs in the air when they
land—can be found throughout North America, they need a
meal already carrying malarial plasmodia to get their chain of
horrors going. Historical evidence supports the arrival of such
microbes in the New World with Spanish explorers and African
slaves, where they became a major problem for settlers in warm,
wet environments like the Chesapeake Bay and Mississippi River
valley. George Washington and Andrew Jackson knew malaria
personally, as did Napoleon. Indian tribes in these and many
other regions were decimated by it. In the first decade of the
nineteenth century the Illinois territory was likened to a grave-

yard because of malaria. The South was especially ravaged by the disease: Divisions of the Union Army fighting below the Mason-Dixon Line during the Civil War suffered 1,213,814 cases, including General Ulysses S. Grant.

Today the disease is rare in the United States and western Europe, where drugs and insecticides (especially DDT after World War II) have largely obliterated domestic germs. Case rates in the United States fell from 34.7 per 100,000 population in 1946 to 1.4 in 1950. But travelers from malarial countries can still start local outbreaks, as long as lady *Anopheles* is around to help. In 1993, three people in New York City were diagnosed with malaria apparently caught from local mosquitoes.

The cycle by which malaria germs thrive and spread is one of nature's darksome wonders. Whenever the right kind of mosquito bites a person with malaria, she sucks blood containing *Plasmodium* cells called gametocytes, which will reproduce sexually (there are male and female types) only inside this insect and nowhere else. The offspring, which are asexual and known as sporozoites, then get injected into any human unlucky enough to be the mosquito's next meal. They congregate in this victim's liver and start multiplying into a form called merozoites. An average of two to four weeks later, the merozoites begin attacking red blood cells, where they feed on hemoglobin and multiply into more merozoites plus a smaller number of new gametocytes. Unless another mosquito comes along at this point, these gametocytes will die and the whole process will reach a dead end—literally, because the destruction of red blood cells has released a toxin that brings on the paroxysmal chills and fevers of malaria; the most prevalent and deadly malarial bug, *P. falciparum*, may cause fevers as high as 110°F. Thus, although a reservoir of bad protozoans (in infected humans) is needed to get an outbreak going, epidemics won't happen without lots of mosquitoes. Note: Malaria can be transmitted by transfusion as long as those sporozoites stay in the bloodstream; one *Plasmodium* species' sporozoites can remain viable for more than forty years.

Obviously there is a chicken-egg problem here regarding how malaria came into being: Who was the first *Plasmodium* carrier, man or mosquito? How did the crazy system of 'zoites and 'cytes evolve? Never mind. By now the cycle is a great mandala of misery in many parts of the world, especially since both the mosquitoes and the protozoans have developed resistance to various insecticides and drugs. In the primordial West African breeding grounds of malaria, natives adapted eons ago to the threat with a gene that produces sickle-shaped, instead of round, red blood cells that are harder for malarial plasmodia to infest. Biologists calculate that the risk of death from malaria must have been at least 25 percent among people without the mutation. It was a mixed blessing, however: Children who inherit this gene from both parents tend to die young. About 0.3 percent of African-Americans suffer from sickle-cell disease.

As early as 1630, the corregidor, or Spanish magistrate, of Loxa, Peru, was relieved of malarial fevers by drinking tea made from the bark of the cinchona (his wife was the countess Chinchon) or "quinaquina" tree, a balm no doubt used by Amerindians for ages. This remedy was a monumental discovery, considering that the standard method of treatment among European doctors included bleeding the victim, whose red blood cells were already being annihilated. Jesuit priests spread knowledge of the bark's power to Europe, where it gained acceptance, at least among Catholics, after use by Louis XIV. In 1658, Oliver Cromwell chose death over use of "Jesuit powder" to cure his chronic malaria. "Much are we beholden to physicians who only prescribe the bark of the quinquina, when they might oblige their patients to swallow the whole tree," remarked the English wit David Dalrymple in 1785.

The medically useful agent was extracted by French chemists in 1820 and called quinine, one of two active ingredients in innumerable gin-and-tonics quaffed on sultry tropical evenings by white imperialists. With the establishment of Dutch cinchona plantations in Java in the 1850s, Europeans could proceed with their empire building in vast parts of Africa previously guarded by *Plasmodium* vectors. The Panama Canal was made possible by quinine and a massive antimosquito campaign in a zone that had been notorious for malaria. Synthetic drugs—derived from dyes developed in Germany between the world wars—were quickly produced after the Japanese conquered Java in 1942 (some half a million cases of malaria occurred among U.S. forces during World War II) and have replaced natural quinine.

The eradication of malaria from Sardinia in 1951 raised hopes for wiping it out everywhere. Between 1955 and 1965, the World Health Organization oversaw the expenditure of $1.4 billion to conquer the disease, but failed. In 1975, Europe was declared free of malaria for the first time in history, yet at that time the disease's worldwide incidence was about 2.5 times greater than it had been fifteen years earlier, having leapt from

"The War with Malaria," a Soviet educational poster.

1 million to 9 million cases in China and from 1 million to 6 million in India.

Today, malaria is becoming more dangerous as plasmodia grow resistant to the man-made drugs. In 1987, all of the Peace Corps workers in the West African nation of Benin came down with the disease while taking chloroquine, the standard penny-a-dose antimalarial medicine for four decades. Vaccines remain an area of intensive research, but none is on the near horizon. In many of the prime malarial regions of the world, the disease is considered out of control.

• pneumococcus •

Lots of microorganisms can cause pneumonia, but the "pneumococcus" is the chief creature behind lobar pneumonia, which damages an entire lobe of the lungs rather than small patches. Identifying it precisely has been something of a Holy Grail mission for researchers over the years, with more than a dozen names being tried until *Streptococcus pneumoniae* was settled upon by the early 1970s. Its popular monikers are far more revealing of its hideous potential than its Latin name: "captain of the men of death," and "friend of the aged."

Louis Pasteur first isolated the pneumococcus bacterium in 1881 from the saliva of a baby who had died of rabies. Investigators subsequently found more than eighty types of pneumococci as they tried to develop vaccines (gold and diamond miners in South Africa, who were dying like cattle under horrendous working conditions, were early recipients of only partly effective immunizations). Laborers on the Panama Canal, especially blacks, fell in great numbers to the pneumococcus. During World War I, pneumonia, fueled by flu and measles but nonetheless due most often to the pneumococcus, was the top cause of death in

military camps. Modern vaccines now prevent widespread outbreaks in high-risk occupations like mining or the military.

Today the primary concern about *S. pneumoniae* is drug resistance, a disastrous phenomenon that could take us back to the days
of Pasteur in a hurry. The prevalence of pneumococcal strains
highly resistant to penicillin increased by a factor of sixty between
the late 1970s and early '90s. In 2000, the Food and Drug Administration licensed a vaccine for infants and young children.

• polio •

Perhaps no other disease illustrates the parochialism of the
developed world more than polio. Since the introduction of
injected Salk vaccine, made from killed viruses, in 1955 and oral
Sabin vaccine, made from attenuated live viruses, in 1960, paralytic poliomyelitis (from the Greek words *polio,* meaning gray, and
myelon, meaning marrow, to indicate inflammation of the spinal
cord) has become a relic of the past wherever immunization of
children is routine. Certainly few Americans under the age of
thirty have ever been haunted by images of iron lungs, steel
braces, or aluminum crutches, let alone carried the disease on
their personal worry list. Yet thousands of cases and deaths still

Board of Health quarantine notice, San Francisco, circa 1910.

occur every year around the world, mostly in Asia and Africa, where immunization is anything but routine.

There are three distinct types of polioviruses of varying potency, and they are the most dangerous members of the entero-virus group. They are bullets of Zen-like simplicity, consisting of little more than a single strand of twelve genes targeted at their host's nerve cells. Polioviruses have been around since ancient times but did not gain much attention until an epidemic hit the United States during the summer of 1916. Though they never caused the massive fatalities of smallpox or yellow fever, the mystery of how they passed from victim to victim and the sud-denness of their crippling attack on robust people spawned a level of superstition and phobia in otherwise "advanced" societies perhaps approached only by HIV in the twentieth century.

Like other enteroviruses, they like to inhabit the alimentary canal. But not just any alimentary canal will do: Only humans, monkeys, and chimpanzees are known to get polio, and humans are the only natural reservoir—a fact that made vaccine research extra laborious, since scientists would much rather experiment with pliant little creatures like mice or guinea pigs. Polioviruses also affect far more people benignly (95 percent of all infections) than with obvious symptoms of disease. Until this fact was fully understood in the 1940s, one of the great puzzles of polio was why such a contagious germ caused relatively few incidents of serious illness—an average of about 16,000 cases of paralytic polio and 1,900 deaths occurred every year in the United States just before the Salk vaccine became available, out of a total pop-ulation of around 150 million. Vast numbers of hidden, but still communicable, infections also explained why quarantining known cases was an ineffective way to halt epidemics.

In the prevaccine era, polio scares were a phenomenon of the well-off industrialized world. As an enterovirus, polio is passed among people mostly via feces from infected individuals, which means that people living in societies with excellent hygiene are more likely to grow up without having their first, naturally immunizing encounter with polioviruses at an early age, when serious disease is a less likely consequence. Scientists even used to think that the poliovirus did not stalk the tropics, because paralysis was so rare in poor equatorial countries. Actually, infections did abound but were asymptomatic, with deaths from "infantile paralysis" getting lost among already high child mortality rates. The prime age for polio in the United States by the 1950s was five to nine years old, with two-thirds of deaths in victims over fifteen. Polio became notorious for striking the "best" families, Franklin D. Roosevelt being the most illustrious example. It was a dangerous side effect, so to speak, of affluence.

Killed-virus vaccine like that pioneered by Jonas Salk was largely supplanted in the United States by the cheaper (pennies a dose), more effective "live" vaccine for universal immunizations. The United States suffered its last documented indigenous case of "wild" poliovirus transmission in 1979 during an outbreak among the Amish in the Midwest (the virus apparently came in from the Netherlands). Risk of domestic exposure to wild polioviruses (as opposed to the tamed viruses in vaccine) is now so negligible that adults are not immunized. Because live vaccine contains a weakened type of poliovirus that can in theory mutate into more virulent forms—an exceedingly rare, but not unheard of, catastrophe—it is not given to people with impaired immune systems or to members of their households. Vaccine viruses are rigorously tested by injecting them into monkeys' spines. Still, between 1980 and 1999, 144 cases of vaccine-associated polio occurred in the United States, leading to the federal recommendation in 2000 that only vaccine made from inactivated viruses be used throughout the country.

In 1988, the World Health Organization set a goal of wiping out polio globally by 2000, but the plan fell short because of the vagaries of war and entrenched poverty. In the Western Hemisphere, the program was successful. By the end of 1991, wild poliovirus had been found in only eight people (all children) living on this half of the planet, with person-to-person transmission of the germ apparently confined to Peru. In 1994, the Americas were certified to be polio-free. Nevertheless, an outbreak of vaccine-derived poliovirus during 2000–2001 in Haiti and the Dominican Republic underlined the necessity of maintaining high vaccination rates. As for the rest of Earth, only 1997 cases were reported in 2006. In 2008, parts of just four countries—India, Pakistan, Afghanistan, and Nigeria—remained as stubborn pockets of endemic infection. Because the virus can be easily imported and then spread quickly among any unimmunized people, nations deemed disease-free will remain at risk—especially the children.

Looking not at all shook up, Elvis gets his polio shot.

• Q fever •

This generally nonlethal disease stumped investigators when it was first studied among Australian slaughterhouse workers in the 1930s; the label "Q" stands for "query." It is now well understood to be caused by a rickettsia-like microbe, originally named *Rickettsia burnetii,* after Sir Frank Macfarlane Burnet (1899–1985), the Australian immunologist and Nobel Prize winner who first isolated it in the late thirties, but now called *Coxiella burnetii,* after H. R. Cox, an American who did similar work at the same time. It is transmitted to humans by livestock or, rarely, ticks and is perhaps best known among pathologists as having more nonhuman reservoirs than any other human illness. Dozens upon dozens of different ticks, flies, lice, chiggers, birds, and mammals may carry the germ for Q fever. It just doesn't bother us much—a lucky thing, since only a single organism is needed for full-blown infection, a phenomenon that puts it on the "Category B" list of potential bioterrorist agents considered to be moderately dangerous.

Q fever's symptoms are much like those of typhus or Rocky Mountain spotted fever, only milder and with no rash. Just a handful of cases turn up every year in the United States and northern Europe, but sporadic outbreaks can occur wherever there are lots of cattle and sheep, especially among workers regularly exposed to livestock herds or products like meat carcasses, wool, and hides. *Coxiella burnetii* can survive for a long time in dust, making barns and other indoor livestock areas especially likely to harbor Q fever; in a classic outbreak during World War II, hundreds of American soldiers camping on a farm near Florence, Italy, caught Q fever in a barn where movies were shown. Since it doesn't hurt the animals, nothing much is done about it besides pasteurization of milk, which has many other justifications. A vaccine exists but is not commercially available in the United States.

Though not a mystery, Q fever still deserves its name because of one remarkable issue: How can an organism capable of massively infecting people from a minuscule initial dose *not* cause serious problems?

• rabies •

Rabies has never been a common disease, but it has affected our metaphorical consciousness like few others. Shakespeare compared jealousy to "poisons more deadly than a mad dog's tooth." Sports fans and other raging fanatics are said to be "rabid." Extremely angry people are described as "foaming at the mouth," though this is usually a slight exaggeration. Shouting "Wild dog!" has little dramatic value nowadays but is still a good attention getter. Rabies is an animal disease, so unless we get in the way of sick warm-blooded mammals that decide to bite us, we will never, ever catch it.

Fascination with rabies is certainly nothing new. Mesopotamian codes before 2000 B.C. mention fatal dog bites. The Greek atomist philosopher Democritus (circa 420 B.C.), who traveled widely in Persia and Egypt, wrote about rabies in domestic animals, as did Aristotle a hundred years later. Epidemics (or, properly speaking, epizootics) of wolf rabies are known to have occurred in western Europe in the late thirteenth century. The first outbreak of dog rabies to leave an extensive record happened in Italy in 1708. The disease apparently did not exist in the Western Hemisphere until the mid-eighteenth century, when dogs caught it in Virginia in 1753. Louis Pasteur performed classic experiments with rabbits and dogs that led to the acquisition of an attenuated virus and the first human vaccination of a nine-year-old Alsatian boy named Joseph Meister, who thereby survived a rabid bite. Though rabies has been eliminated in many countries through pet control and careful quarantines, it is endemic in various wild animals in the Americas and in domestic and wild animals throughout Africa and Asia, where most human cases occur today. About 50,000 deaths are attributed to rabies every year, the vast majority in poor countries. Between 1995 and 2006, 37 deaths were reported in the United States, of which 28 cases were associated with bats. Hawaii is the only state that has never reported an indigenous case of human or animal rabies.

The virus that causes rabies is said to be "neurotropic." That is, it seeks out nerve cells. Evolution has pumped such ferocity into this organism's craving for nerve tissue that it literally drives its victims crazy in order to jump to the next host, via the saliva in the rabid animal's bite wounds that break the skin. In fact, the salivary glands actually become infected before most animals show any overt sign of rabies. The disease has a long incubation period, averaging a month or two in humans but sometimes lasting years, depending a lot on the location (head, face, or hands are most dangerous) and severity of infective bites. This provides a window for treatment, which is generally futile after symptoms begin. Professionals are fond of saying that early diagnosis does not alter the patient's prognosis. Rabies victims at first feel a general malaise and restlessness, then grow increasingly agitated, with painful spasms of the throat. Some start to hallucinate. Soon the victim cannot drink, which is why rabies has been called "hydrophobia." Death comes within about ten days after the appearance of symptoms. Dogs actually show two forms of the disease: "furious rabies," the classic state of incoherent viciousness, and "dumb rabies," where paralysis takes over. Wild animals can get "furious," too, but they more often start by showing uncustomary behavior—nocturnal creatures come out in the daytime, the timid grow bold, etc.

In the United States and Europe, where rabies in dogs and cats is largely controlled through vaccination, wild animals such as raccoons, skunks, bats, foxes, and coyotes are the virus's main

A cartoon published in 1826 shows a rabid dog causing panic in the streets of London—for good reason.

reservoirs (humans are a dead end, since they do not transmit the virus and suffer essentially 100 percent fatalities if not treated immediately after getting infected). Of 6,940 animal cases reported in 49 states, the District of Columbia, and Puerto Rico in 2006, wild animals accounted for 92 percent, with raccoons the highest at 37.7 percent, followed by bats (24.4 percent), skunks (21.5 percent), and foxes (6.2 percent). Pets should not be presumed to be rabies-free. In October 1994, 500 people in New Hampshire had to be immunized after exposure to possibly rabid kittens from a Concord pet shop. The kittens—one died and tested positive, twelve others were destroyed and tested negative—may have been infected by a raccoon before delivery to the store. Rabbits and rodents, including squirrels, are almost never infected. Rabies can also be transmitted by eating infected meat, as Darwin observed among Peruvians who ate beef from a rabid bull. The virus may also be airborne, as a few unfortunate explorers of bat caves have discovered.

Most people bitten by a rabid animal will not catch the disease, but if a suspect critter is involved, vaccination should start as soon as possible anyway. There is only one method of testing an animal: Kill it and examine its brain. Washing a bite thoroughly with soap is the best first aid. Though uncommon, the mere threat of rabies is expensive—a confirmed case in one person may make it necessary to vaccinate hundreds of others at a cost of hundreds of thousands of dollars.

• rhinoviruses •

The rhinovirus is the king of colds, aptly named for the havoc it has been wreaking upon human noses since, well, humans have had noses. Rhinoviruses come in so many different types, probably hundreds, that a universal cure is hopeless—you could spend a lifetime catching one new kind after another and never run out, which is exactly what happens. From Pliny the Younger, who in circa A.D. 100 prescribed "kissing the hairy muzzle of a mouse," to Sir Alexander Fleming, who in 1945 recommended "a good gulp of hot whiskey at bedtime—it's not very scientific, but it works," cold remedies have run the gamut of imagination. But the rhinovirus endures.

Rhinoviruses just happen to thrive at the average temperature of our nasal mucous membranes, about 93°F. They can survive as long as two hours on skin and four days on nonporous surfaces. They are cousins of poliovirus, possessing about half the same genes, which may be why they show the same tendency to strike in the late summer and fall (in the Northern Hemisphere, that is). No one has figured out exactly how they spread—close-up exposure to infected snot, pardon the colloquialism, is the best guess, accomplished by carrying germs on the hands and fingers up to the nose and eyes. Several enterprising microbiologists ran an experiment in 1973 that showed kissing ("direct oral contact") to be an inefficient way of spreading rhinoviruses. Other researchers found in 1977 that psychological stress makes rhinovirus colds worse. No one has taken the next leap forward to determine the effects of stressful kissing.

Though it is impossible to tell a rhinovirus cold from any other kind by symptoms alone, noses may get extra runny and coughs are especially common. Rhinoviruses are known to trigger asthma in children. A rhino cold lasts on average about a week, with peak misery on the second and third days. There is shaky evidence—far from conclusive—that heavy doses of vitamin C may reduce the lengths of colds, but not the frequency of catching them. Washing your hands a lot is a better bet.

So the old proverbs are as good as any other remedies. From England: Stuff a cold and starve a fever. From Germany: Sauerkraut is good. And from India, where they should know: One cold in the head is as bad as ten diseases. Keep in mind this useful rule of thumb: Untreated colds last a week; medical attention can end them after seven days.

• *Rickettsia* •

Creatures in the genus *Rickettsia* occupy a niche between bacteria and viruses. They carry much of their own cellular equipment and are vulnerable to antibiotics, but like viruses they need to invade living cells in order to grow. They were named after Howard Taylor Ricketts (1871–1910), an American pathologist who identified them as a family in 1907 and later died from typhus in Mexico.

Most rickettsial diseases fester among animals over the long term and reach humans via insects that carry the microbes. Rocky Mountain spotted fever is caused by *Rickettsia rickettsii*, which is transmitted by ticks. Typhus comes from *R. prowazekii* carried by body lice or rat fleas. Rickettsialpox, a mild fever-rash disease first seen in New York City, is carried by mouse mites infected with *R. akari.* Scrub typhus, found across the region of Japan, India, and Australia, comes from *R. tsutsugamushi* and is transmitted by chiggers from various rodents. Myriad types of tick-borne typhus are known simply by where they happen: Queensland, Marseilles, India, etc. In general, the Eastern Hemisphere's tick-borne diseases are milder than the Western Hemisphere's.

Because of the involvement of insects often associated with poor hygiene, rickettsial diseases used to mean serious problems for the military. Trench fever, a now-rare illness that results when infected louse feces containing *R. Quintana* (reclassified as *Bartonella quintana,* thus closely related to *Bartonella henselae,* which causes cat scratch disease) are rubbed into itchy skin, got its name from the miserable battlefield conditions of World War I, though it plagued Napoleon's armies long before then. Not much was known about it until the hellish confrontations in Flanders and France in 1915. Between 1915 and 1918 it accounted for one-fifth to one-third of all illness in British forces in France and about one-fifth in the German and Austrian armies. By the time American soldiers arrived en masse on the Continent, the louse's role was understood, thanks to volunteers who allowed themselves to be fed upon by body lice from trench fever patients. The disease is still endemic in Poland and Russia, as well as in Mexico and northern Africa. So-called urban trench fever is the disease's modern devil among homeless alcoholics in the United States. *Ehrlichia chaffeensis,* a rickettsial bacterium first recognized in 1986, causes systemic infection, with fever, headache, a low white-blood-cell count, and sometimes meningitis in people bitten by the lone star tick, *Amblyomma americanum.* It emerged in the southern and mid-Atlantic states. Nature never rests.

• Rocky Mountain spotted fever •

This is one of those uncommon diseases, like rabies, that have a way of generating more than their fair share of fear. Over the years, from 250 to 1,200 cases have turned up annually in the United States, with fatalities of about 7 percent in the most vulnerable patients, provided antibiotic treatment is started promptly. Yet many a parent will go ballistic at the sight of Junior walking across a summer meadow, because *you might get a tick bite!* Kids should be examined for ticks after a day outside in infested areas, but really, Mom and Dad, let them play.

Rocky Mountain spotted fever is caused by *Rickettsia rickettsii,* an organism somewhere between viruses and bacteria on the evolutionary ladder. It is found naturally only in the Western Hemisphere (other rickettsia cause various tick fevers in the Eastern Hemisphere). Humans are accidental hosts and have no role in sustaining *R. rickettsii,* which depend on certain Ixodidae, or hard-backed, ticks. They include dog ticks (*Dermacentor variabilis*), which are ubiquitous east of the Rockies and most often responsible for infecting humans with *R. rickettsii*; wood ticks (*D. andersoni*), found in the Rocky Mountain states; but *not* the deer ticks of Lyme disease fame. The name Rocky Mountain is somewhat misleading: Though the disease was first recognized in those Western states, it is relatively uncommon there and has been encountered in all the lower forty-eight except Maine, with highest frequency in the Southeast. Even in areas where most cases occur, only about 1 to 3 percent of the tick population carries *R. rickettsii.*

The earliest reports of the affliction came from doctors in Idaho's Snake River region in the 1870s, who called it "black measles" because of the horrendous rash it brings in late stages. Folk wisdom ascribed it to drinking melted snow or breathing sawdust. The Rocky Mountain Laboratory (now part of the

National Institutes of Health) was established in Hamilton, Montana, to study the alarming disease. A physician in Great Falls, Montana, with the perfect name of Earl Strain, first perceived an association of the symptoms with tick bites in 1902. Then the illustrious young pathologist Howard T. Ricketts took up the subject at the University of Chicago between 1906 and 1909, describing scientifically the tick vector and the process of microbe transmission. In 1909, the Montana legislature voted $6,000 for his crucial research, but the state's board of examiners never sent him the money. Ricketts left for Mexico to study typhus in July of that year, where it killed him in 1910 at the age of thirty-nine. To honor his memory, the organisms that cause typhus and sundry spotted fevers were dubbed rickettsia in 1916.

Before antibiotics cured most cases of Rocky Mountain spotted fever starting in the late 1940s, mortality averaged about 20 percent (lower in children, higher in heavily exposed older adults). After an incubation period of three to twelve days, the victim gets a sudden, splitting headache along with high fever, nausea, and chills. After about four days of fever a pink rash appears on the wrists and ankles that quickly spreads from extremities to the trunk and gets darker. (Oddly enough, typhus rash spreads the other way, from trunk to extremities.) If untreated, severe infections lead to pneumonia, shock, and death. There is no vaccine licensed for general use. The disease is not spread from person to person; it requires a bite from an infected tick or exposure to tick feces (hence the warning not to crush a tick when removing it from the skin). *R. rickettsii* often require a meal of fresh blood to become activated into a virulent form, which means the tick must stay attached to a person for several hours in order for the disease to be transmitted.

Regarding the perpetually contentious subject of tick removal in the field: Nothing beats a plain old pair of fine-tipped tweezers, followed by disinfectant. A hot match on the tick's butt may just encourage the little bastard to upchuck more bacteria. Remember, a slug of good whiskey helps get the patient through the operation—more than enough reason to carry a flask in your backpack.

• rotavirus/Norwalk •

About half of all babies and young kids hospitalized with diarrhea are infected with rotaviruses (from the Latin *rota,* meaning wheel, referring to their characteristic shape), which mess up the insides of many animals, too. Along with the so-called Norwalk (Ohio) viruses and related caliciviruses, rotaviruses play a monumental role in human health all over the planet, where there are billions of cases of diarrhea every year and, in poor countries, millions of resulting deaths.

Both types of viruses were discovered in the early 1970s. Before then, the pathogens behind most cases of gastroenteritis ("stomach flu," not to be confused with influenza) and epidemic diarrhea in young children were a mystery; bacteria known to cause diarrhea, such as certain pathogenic strains of *Escherichia coli,* are more prevalent in older children and adults. The Norwalk viruses and rotaviruses pass around through infected feces. They are therefore classic "filth" agents, turning up in drinking water, swimming areas, uncooked foods (especially shellfish), and along other fecal-oral routes wherever sanitation is awful. A rotavirus vaccine for children called Rotashield was approved in the United States in 1998, but it was pulled from the market the following year because of safety concerns. The currently licensed oral vaccine Rotateq is given to healthy babies in several doses within the first eight months of life.

• *Salmonella* •

The leading cause of food poisoning, non-typhoid *Salmonella* bacteria (which refers to strains that *don't* cause typhoid) have surged in medical importance during the past several decades, due to drug-resistant strains and lax food handling. In the United States and several European countries, raw or undercooked eggs are a prime source of *Salmonella enteritidis,* though only about one egg in every ten thousand carries the germ in the northeastern United States. In 1999, there were forty-four confirmed outbreaks in seventeen states, producing 1,080 sick people and sixty-three hospitalizations. But perhaps 1.5 million Americans are actually made ill every year by various *Salmonella* bugs, and more than 500 die. Even for many baby boomers who have never been sick, *Salmonella* holds a rancid place in modern history as the reason they had to get rid of their little green pet turtles in early childhood. Some of them may also remember when marijuana was rumored to be contaminated.

Salmonella poisoning is not especially dangerous as long as one's immune system is working properly or effective antibiotics are available for complicated cases. Gastroenteritis is the technical name for nausea and cramps that start twelve to seventy-two hours after ingestion of the bacteria, soon followed by diarrhea, fever, and maybe vomiting. The whole mess should go away in four to seven days, leaving the victim adamant about never eating again whatever was at fault.

Salmonella is a very big genus, with twenty-two hundred known members. (The surname, so to speak, comes from the American veterinarian Daniel E. Salmon, who first described the pests in 1885.) Some prefer humans, some don't, most are ubiquitous. The more common types are named after the place where they first spoiled a party: *S. heidelberg, S. newport, S. monte-*

video, S. saint paul. One has an especially apt moniker: *S. agona.* The U.S. government established meat inspection laws in 1906 to prevent infections, but the record since then has been far from spotless. As new ways to get sick are discovered, new regulations are written; for example, milk and eggs that will be dried for myriad food products must first be pasteurized—a lesson learned the hard way, of course. In October 1994, thousands of Americans fell ill after eating contaminated ice cream from a Schwan's plant in Minnesota, leading quickly to new rules about when certain preliminary ingredients need to be pasteurized and the use of tanker trucks. The incident resulted in at least 489 confirmed cases in twenty-nine states, with another 3,342 probable or suspect cases in thirty-nine states. Any food that contains raw eggs is suspect—cookie batter, hollandaise sauce, Caesar salad, etc. Vegetarians are not immune: Since 1995, fifteen outbreaks have been associated with eating raw sprouts.

Multiple resistance to antibiotics has become a big bugaboo with *Salmonella,* especially since they can transfer their resistance to other gastrointestinal bacteria, such as *Escherichia coli.* Overabundant use of livestock and chicken feed laced with antibiotics is a primary force behind the evolution of drug-resistant strains that then find their way into people. Anybody already taking antibiotics for something rather ordinary, like an earache, is in for a bad ride if he eats a burger contaminated with *Salmonella,* since the very likely drug-resistant food bugs will land in territory that has been swept of competition.

So mind the beef—have a nice, carefully washed salad with oil and vinegar. At home, wash hands, utensils, and surfaces carefully with hot, soapy water after preparing raw chicken—more than one bird in three is probably infected. And don't drink anything containing raw eggs, no matter how bad your hangover is.

• San Joaquin fever •
(Coccidioides immitis)

Inhaling the spores of the soil-dwelling fungus *Coccidioides immitis* can cause the illness known as San Joaquin fever. Americans have to travel to dusty places in the Southwest to have any chance of catching it. Many infected people notice nothing, or maybe flu-like symptoms at most. Some will get a romantic-sounding but no doubt bummer affliction known as desert rheumatism. They'll all probably be okay with no treatment. But anyone with a weak immune system may—weeks, months, or years later—come down with an increasingly awful set of problems in his lymph nodes, bones, liver, brain, and other organs. Nearly two-thirds of them will die. Progressive coccidioidomycosis, as this curse is called technically, is one of the defining diseases for AIDS.

• slapped cheek disease •
(parvovirus B19)

The human parvovirus B19 causes a mild red "slapped cheek" rash on the face of an infected child after several days of common cold symptoms. Slapped cheek disease is hardly as fearsome as getting your cheek slapped for real, but, rarely, kids with sickle-cell anemia may become seriously ill, and pregnant women may incur damage to their fetus or even suffer a stillbirth. The condition is also known as fifth disease, referring to its place among the six classic rash illnesses of childhood: measles, scarlet fever, rubella, something called "Duke's disease" (never clearly distinguished by the old-time docs), fifth, and roseola. Child protection workers should be acquainted with it.

• smallpox •

Smallpox was the least understood and most destructive disease in history, yet the first to be eradicated from nature. It was caused by a fiercely infectious germ of the *Orthopoxvirus* genus, which includes a menagerie of other plagues: cowpox, rabbitpox, buffalopox, monkeypox, camelpox, raccoonpox, et cetera (but not chicken pox). One droplet of exhaled moisture from the lungs of an infected individual would contain enough smallpox viruses to infect a thousand other people. The fact that we don't have to worry about this devil anymore shows what can be done with enough money. However, the fact that it would obviously make a fabulous bioterrorist weapon has brought it back like Dracula to haunt the minds of public health specialists, security officials, and anyone else who contemplates such horrors.

Between its first recorded attack in ancient Egypt and the last in 1977, smallpox claimed hundreds of millions of victims. Telltale signs of a smallpox rash on the mummy of Ramses V (who died in 1160 B.C.) indicate the disease's antiquity. The Roman Empire lost more than a third of its subjects in some areas during a fifteen-year epidemic that began in A.D. 165. Crusaders brought smallpox back with them from the Holy Land—more than sufficient revenge by the infidels. Conquistadors carried it to the New World, where it mowed down the Aztec and Inca empires. The word "smallpox" began to be used in English during the sixteenth century as the translation of the French *la petite vérole (vérole = pox); la grosse vérole* was syphilis. In the American colonies smallpox was a primary force allowing Europeans, who had developed immunities, to eradicate the culture of indigenous peoples, who had not. Pocahontas died of smallpox in 1617 after visiting London. In Europe it maintained the power to strike all levels of society, killing Queen Mary II of England in 1694, Emperor Joseph I of Austria in 1711, and King Louis

Sopona, god of smallpox—a wooden statue adorned with hair, shells, beads, and a sheaf of plants—worshiped by the Yorubas of Nigeria.

XV of France in 1774. In Japan it claimed Emperor Gokwomyo in 1654 and Emperor Kōmei in 1867. Isolated regions could be devastated, as when more than a third of the population of Iceland perished in 1707.

The quality of professional medical care during this era may be surmised from the record left by a smallpox patient of Thomas Sydenham (1624–1689), one of the most illustrious English physicians: "In the beginning I lost twenty ounces of blood. He gave me a vomit, but I find by experience purging much better. I went abroad, by his direction, until I was blind, and then took to my bed. I had no fire allowed in my room, my windows were constantly open, my bedclothes were ordered to be layed no higher than my waist. He made me take twelve bottles of small beer, acidulated with spirit of vitriol, every twenty-four hours. I had of this anomalous kind to a very great degree, yet never lost my senses one moment."

Shitala Mata, Indian goddess of smallpox, sitting sidesaddle on a horse while holding a pot and a sheaf of fennel, perhaps on her way to a nasty business conference with Sopona.

In 1714, a Greek physician named Emanuel Timoni published an article about how he had prevented the disease during an epidemic by embedding a knife in a victim's rash, then scratching a healthy person with the blade. This method, which must have seemed like pure magic, was apparently similar to age-old folk techniques known across China, India, and western Asia. In 1721, Lady Mary Wortley Montagu, the ebullient wife of the English ambassador to Turkey, stunned polite circles with the story of how she had permitted a Gypsy woman to immunize her infant son in a similar way. ("I am patriot enough to

The spirited Lady Mary Wortley Montagu (1689–1762), by all accounts magnificent in mind and body.

take pains to bring this useful invention into fashion in England," she had written to a friend in 1717, "and I should not fail to write some of our doctors very particularly if I knew any one of them that I thought had virtue enough to destroy such a considerable branch of their revenue for the good of mankind.") Children of the English royal family were then immunized after the king first tested the procedure on six condemned prisoners and eleven charity children.

That same year, Cotton Mather, who had heard of the practice from his slave, convinced fellow Congregational ministers to advocate such a desperate measure, as smallpox was threatening to depopulate Boston. Some 6,000 residents of the city contracted the disease, and 900 of them died. Of 287 who braved primitive immunization, only 6 perished. Despite this success, inoculations were widely illegal in the American colonies, mainly because the often slapdash technique itself could start epidemics and was poorly explained to an understandably wary public (indeed, there was then no real explanation available). Benjamin Franklin, who greatly mistrusted the medical expertise of his day, could not rise above the counterintuitive proposition of exposing someone to a ghastly disease in order to save them from it—his four-year-old son died in a 1736 epidemic. John Quincy Adams, on the other hand, was inoculated at the age of eight in 1775. Educated Europeans also had trepidations: Mozart caught smallpox, which he survived, because his father believed that preventive action interfered with God's will. But Catherine the Great of Russia, prodded by Voltaire, paid an English doctor £10,000 plus a lifetime annuity of £500—a fee that even modern physicians might envy—to inoculate her and the Imperial Court. Immanuel Kant thought the procedure was deadly, however.

Only after George Washington ordered his army to line up during the Revolutionary War were anti-inoculation laws ended. Still, smallpox felled at least 130,000 Americans during the con-

A drawing from "An inquiry into the causes and effects of the variolae vaccine," published in London in 1798, showing the milkmaid's hand Jenner used as a source for his vaccine.

flict. The Boston Tea Party might have been a smallpox massacre had not the infected crew of the brigantine *Beaver* been quarantined in December 1772 before joining other ships at Griffin's Wharf. General Washington, who had survived a bout with the disease after visiting Barbados in 1751 but was severely scarred by it, ordered that mail from Boston be doused with vinegar to kill the germs. Harking back to when the British had conducted biowarfare with smallpox against the Indians at Fort Pitt in 1763, the press accused Cornwallis of trying to spread the disease among New Yorkers.

In 1796, an English country doctor named Edward Jenner found a less threatening procedure that became one of the great milestones of preventive medicine. He had noticed that milkmaids who caught cowpox, a similar but relatively mild disease transmitted by cows, were somehow immune to smallpox. He therefore tried injecting an eight-year-old boy named James Phipps with fluid or "vaccinia" (hence the word "vaccine") from a cowpox pustule and, six weeks later, pus from a smallpox lesion. The brave boy reacted slightly to the cowpox inoculation but did not catch smallpox at all. Though the Royal Society rejected Jenner's paper describing his groundbreaking experi-

An illustration published in 1802 by a London antivaccination society shows Edward Jenner inoculating a young woman as former patients suffer the side effects of his vaccine, which consisted of fluid from a cowpox pustule.

ment and twelve subsequent trials, he soon published a book that provoked widespread testing to confirm "vaccination" as safe. It was soon in general use, and Parliament thanked Jenner with an award worth a then colossal $150,000. "You have erased from the calendar of human afflictions one of its greatest," Thomas Jefferson wrote to Jenner in 1806.

Still, smallpox remained a serious problem for more than a century outside the sphere of Jenner's vaccine, including in the United States, where massive immigration and antivaccination societies kept health officials on the defensive. "A girl cannot be termed beautiful until she has had smallpox," pronounced a Sicilian proverb, reflecting the disease's ubiquity. Abraham Lincoln fell ill with smallpox several hours after delivering the Gettysburg Address. By the 1950s, compulsory vaccination of schoolchildren had largely freed North America, Central America, and Europe of the scourge. Hospitals were among the last major wellsprings of infection.

In other parts of the planet, however, there were still 10 to 20 million cases of smallpox per year, causing 2 million deaths. The disease was so common in Bangladesh that the word for springtime, *bashunto,* was also the term for smallpox. In 1975, the last incidence of wild variola major, the most virulent form of the disease, was cured in a three-year-old Bangladeshi girl named Rahima Banu. Finally, in 1977, after a $330 million, ten-year campaign (first proposed by the Soviet Union in 1958) supported by more than 150,000 workers in forty-one African, Asian, and South American countries, smallpox viruses were confined to research labs. The last natural case of smallpox, variola minor, occurred that year in a Somalian hospital worker named Ali Maow Maalin.

Today the last remaining germs are frozen with liquid nitrogen in about six hundred inch-long test tubes stored under maximum security at the Centers for Disease Control and Prevention in Atlanta and the Center for Research on Virology and Biotechnology in Koltsovo, Russia. These prisoners were slated for execution on New Year's Eve, 1993, but won a reprieve after some scientists insisted they could be valuable for future study of the human immune system. Knowing that the infamous *Orthopoxvirus* might become the first species to be made deliberately extinct, researchers drew a map of its DNA sequence for future reference. In 1999, the World Health Organization recommended that the lab stocks be destroyed by the end of 2002, but they postponed that action indefinitely after the bioterrorism of 2001. Death by pressure cooker at 248°F for two forty-five-minute cycles is still a matter of debate between those who think it's safe to get rid of them and others who believe they might still yield useful information.

Fear that germs might accidently escape—in 1978, a medical photographer died from smallpox after working one floor above a lab in Birmingham, England (the resident virologist later commited suicide)—has thus been put aside while the World Health Organization considers their fate and vaccine research continues. Surveillance for even a single new case of the disease remains active, especially where infection might come from animals via mutated *Orthopoxvirus*. Russian scientists even dug up smallpox victims who had been buried in Siberian permafrost to see if the virus was present in their flesh; it was not. Some health workers and military personnel are still vaccinated against the disease. Otherwise, it's long gone.

Which is why smallpox is perhaps *the* most worrisome bioterrorist nightmare. Vaccination of the general public ended in the United States in 1972 and the rest of the world in 1980. Thus, a single "smallpox martyr" infected from secret or stolen stocks of the virus could start a global holocaust (death rate: 30 percent) in populations without any immunity during the seven to seventeen days before he or she showed symptoms of the disease. As a precaution, the Centers for Disease Control and Prevention has for years stored some 8.5 million doses of vaccine (which are not commercially available because of medical hazards associated with their use). In 2000, the U.S. government awarded a contract to start producing 300 million new doses.

• *Staphylococcus* •

In 1928, the British scientist Alexander Fleming found by accident that *Penicillium* mold could kill the dreaded *Staphylococcus* bacterium, thus launching the golden age of antibiotics. Fleming gave up on trying to isolate the specific substance involved, but scientists at Oxford succeeded in 1940 in isolating penicillin. During World War II, development moved to the United States, where large-scale production and testing of penicillin took place—with the rather unfair twist that the Brits had to pay royalties to manufacture the miracle drug they discovered. Penicillin was truly a wonder: potent against pathogens in tiny amounts, safe to inject directly into the bloodstream, showing low toxicity even in high doses. Understandably, it became a universal lifesaver, but its very success led to an evolutionary backlash: The pathogens learned how to beat it. By 1982, penicillin could stop fewer than 10 percent of *Staphylococcus* (staph) infections. There were alternative drugs, of course, but by the 1990s the widespread existence of super-resistant *Staphylococcus* strains was a genuine crisis, especially in poor countries where the few remaining effective antibiotics are impossibly expensive.

Staph infections are no doubt primordial. The boils that Job scraped with broken pottery could only have been caused by staph. But the bacteria were not observed and studied until the late nineteenth century, when Sir Alexander Ogston of the Royal Aberdeen Infirmary named the organisms after the Greek words for grape (*staphyle*) and berry (*kokkos*) They grow like bunches of grapes everywhere. From 30 to 50 percent of healthy adults carry them harmlessly in their noses, and 20 percent have them on their skin. There are two species: *S. aureus* and *S. epidermidis* (originally called *S. albus* to differentiate its white colonies from *aureus*'s golden orange colonies). Most infections are confined to the skin and are not serious, let alone fatal. But if they spread and enter the bloodstream, the consequences can be dire. They are particularly prevalent in hospitals—of about 23 million surgeries performed in the United States in 1992, 920,000 resulted in bacterial infections, most of which were staph, a rate rivaled only by *Escherichia coli*.

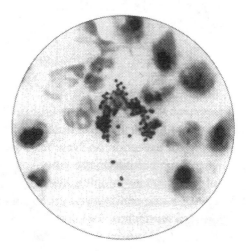

Staphylococcus *bacteria, bunched together like grapes, magnified a thousand times.*

S. aureus, the more virulent of the two, produces a powerful toxin that can cause food poisoning, which is felt very quickly, in as few as thirty minutes after eating, compared to other stomach turners that take hours to kick in. Chinese mushrooms were taken off the U.S. market in 1989 when it was discovered that transporting them in plastic bags could help cultivate *S. aureus.* The microbe is highly contagious through contact with its colonies in wounds, and it can bring on pneumonia. It is also the devil behind toxic shock syndrome, which came to public attention in the early 1980s among women who used super-absorbent tampons. The appearance of *S. aureus* bacteria with reduced susceptibility to vancomycin, a last-resort antibiotic, is of great concern.

The human immune system is very good at combating most staph infections, so *Staphylococcus* preys hard on the weak: infants, the elderly, the malnourished, drug abusers (especially heroin addicts), the immune-impaired, the obese. Burn victims and flu sufferers are known to be more susceptible, as is anyone with foreign objects in his or her body (sutures, artificial valves or joints, intravenous tubes, tampons, etc.). But because of their ubiquity, *Staphylococcus* germs can strike anyone given half a chance. George Bush experienced a severe staph infection in his right arm as a teenager, though it did not prevent his raising that arm to take the oath of office later in life.

• *Streptococcus* •

"Getting your tonsils out" used to be one of childhood's rites of passage—for many people the first and, with luck, the only stay in the hospital until old age. Plus you could have all the Popsicles you wanted to soothe the throat pain afterwards. Influential researchers had concluded from the fact that many kids' tonsils were constantly swollen that this "focus" of *Streptococcus* infection should simply be junked. Tonsillectomy was good for you and even prevented more serious afflictions. This all turned out to be flat wrong, but not before millions were robbed of a perfectly fine part of their anatomy. And it wasn't even a religious ritual.

Streptococcus bacteria are admired for growing in beautiful chains; hence the name, first used in 1874, which means a twisted chain of berries. Strep germs are probably most despised today for causing bad sore throats and, occasionally, scarlet fever (which is just sore throat with a red rash and vomiting). They are common throughout the world in school-age children, many of whom are symptomless carriers rather than sick puppies. Incidence of both strep throat and scarlet fever has dropped dramatically since the beginning of the twentieth century, perhaps because of decreasing *Streptococcus* virulence and/or increasing human immunity. Of course, antibiotics now keep the bacteria from multiplying in an individual to life-threatening levels.

The first recorded epidemic of scarlet fever, then called

Twisting chains of Streptococcus *bacteria, magnified a thousand times.*

rossalia, happened in Sicily in 1543. Fluctuating widely in severity, the disease was often confused with measles, as when Samuel Pepys wrote in his diary for November 10, 1664: "My little girle Susan is fallen sick of the meazles, we fear, or, at least, of a scarlett fevour." Scarlet fever often occurred when strep got into the milk supply, perhaps from a dairy farmer with a sore throat, an accident now precluded by pasteurization. (Yet one of the most common microbial causes of harmlessly sour milk is a *Streptococcus.* Another strain gives yogurt a creamy taste.) Before 1830, the disease was apparently mild in America, but deadly epidemics then struck regularly across the country until it was a leading cause of childhood mortality. Fatalities, though not necessarily the number of attacks, began to decline around 1880. By the 1970s, deaths in the United States from once-feared scarlet fever had plummeted from more than six thousand a year in 1910 to a negligible dozen or so. Rheumatic fever, another consequence of strep infection that used to be feared as a predisposer to heart

Pneumococcus cells, aka "friend of the aged," magnified a thousand times.

disease later in life, is now rare in the United States, though sporadic outbreaks have been trending upward since 1985. It is still awfully common in poor countries.

Crowded living conditions help spread strep through close contact, as classically demonstrated during the American mobilization for World War II, when Navy recruits suffered 59,458 cases of scarlet fever and 43,448 cases of rheumatic fever (other military services had similar outbreaks). During World War I, strep infection was the main complication of influenza leading to fatal pneumonia. Today, *Streptococcus pneumoniae,* also called the pneumococcus, is the most common cause of bacterial pneumonia around the world; it is the germ behind as many as half of all ear infections, the most frequent reason for taking kids to the doctor in the United States (about 24.5 million miserable trips every year). Up to 40 percent of these infections are resistant to one or more common antibiotics. In the absence of earaches, pediatricians might need to moonlight by delivering pizzas or driving taxis from time to time, just like humanities Ph.D.s.

Unlike the ever-mutating *Staphylococcus* bug with which it is often mentioned, strep did not raise serious concern about resistance to antibiotics until the 1970s, when penicillin-resistant strains began to appear. As so-called strep-A bacteria, historically the most common and virulent type in humans, seemed virtually to disappear, strep-B strains rose to the occasion, particularly among infants. Then strep-A reemerged in the late 1980s stronger than ever. A highly resistant *S. pneumoniae* was part of this new microbial offensive, shocking the American public when it suddenly took the life of beloved Muppets creator Jim Henson in May 1990 at the age of fifty. Around the same time, a thirty-year-old Canadian man died six weeks after *S. pneumoniae* infected a splinter cut on his finger. Between the 1940s and 1990s, the typical dose of penicillin required to fight strep respiratory infections got bigger by a factor of 2,400 due to tougher bacterial resistance. In 1993, pediatricians in Kentucky reported that 28 percent of *S. pneumoniae* cultures from children with acute ear infections were resistant to penicillin.

Strep-A's potential to cause severe skin infections was sensationalized in the United Kingdom in the spring of 1994 when press reports circulated of a deadly "flesh-eating" bacterium that emits an enzyme able to destroy the fascia, or tissue that binds skin to muscle. The germ's invasive speed is unmatched by any other infectious organism. One Tory MP from Gloucester, where a cluster of cases occurred, called the bug "galloping gangrene that can eat flesh at the rate of an inch an hour." Headlines such as "Public told not to panic as bacterium death toll rises to 12" in the *Times* of London no doubt helped sell the story. Level-headed physicians eventually assured the public that the situation was not unprecedented, and rare in any case. About six hundred cases of "necrotizing fasciitis" occurred in the United States in 1999, with fatalities around 20 percent in otherwise healthy adults. In November 1994, the fifty-year-old chancellor of a California community college district died from such an infection five days after complaining of sore throat and leg pains.

Maybe some enterprising editor looked up the term "necrotizing fasciitis" and decided "flesh-eating" would be more likely to awaken the morning commuters. No doubt the faxes between London and Hollywood were quite active for a while, if not the medical hot lines. Serious streptococcal disease does seem to be on the rise, however, perhaps because invasive strains are circulating against which we have not had time to develop much "herd immunity," as epidemiologists say.

• syphilis •
(Treponema pallidum)

The pox. The serpentine malady. Hot humours from foul women (and vice versa). The French disease. Or, depending on one's national allegiance, the Spanish, Italian, German, Polish, or Portuguese disease. Nobody ever wanted to claim syphilis. Lots of people of every stripe caught it. In 1530, during the era of its first and most virulent attack in Europe, an Italian physician named Girolamo Fracastoro, a classmate of Copernicus's at Padua, published a moralistic poem describing its symptoms and how to treat it. For the sake of metaphor he also told the myth of Syphilus, a shepherd who incurred Apollo's wrath and got punished with "foul sores." The tale and the venereal disease resonated deeply with Christian notions of sin, tying poor Syphilus's name forever to the divine retribution that awaits anyone daring to do what comes naturally. If you think Western civilization has advanced past the Middle Ages in this regard, you probably haven't picked up a newspaper or magazine since the AIDS epidemic began.

Fracastoro was no fool. In hypothesizing that syphilis was caused not by a "mysterious shadow or miasma, not by obstructed humors, but by a kind of seed," he was on the brink of a germ theory of disease.

Treponema pallidum ("pale turning thread"), the beautiful spiral bacterium that causes syphilis, can be killed in a few minutes by soap and water. It probably cannot penetrate unbroken skin (shaven skin, maybe). It does not like dry places. The disease is thought of as highly contagious, but being exposed to the bug does not always mean infection. Transmission, when it does happen, is usually by hetero- or homosexual intercourse, or heavy kissing, which allow germs to pass from a moist chancre on the already infected partner's genitals or lips to a moist site on the uninfected object of affection. A new sore develops at this

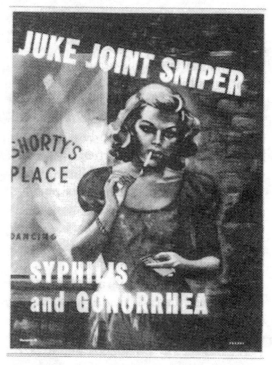

Always one of Shorty's girls, never Shorty.

point of contact, followed weeks later by a generalized rash and fever. Then symptoms disappear during a latency period that may last for years. The infection is quietly spreading, however, causing damage throughout the body. The horrible "tertiary stage" brings arthritis, heart disease, madness, and death. In the final scene of Henrik Ibsen's 1881 play *Ghosts,* young Osvald Alving—a victim of congenital syphilis transmitted from his mother during pregnancy—cannot tell the difference between night and day.

Since the bacteria eventually enter the bloodstream, syphilis can also be passed via transfusions. Penicillin will stop the devil in its tracks, at least in the disease's early stages, but there is no vaccine: *T. pallidum* is so well adapted to living tissue that it cannot be artificially cultured. Luckily, it has never developed resistance to antibiotics.

Long before these basic facts were collected, syphilis cut a swath through society that was more reprobative than deadly. The malady was first noted in Europe at the end of the fifteenth century, though it may have existed before then undistinguished from leprosy or as a disease not transmitted sexually. Archaeologists have recently discovered skulls dating back to 1300 in an English church that show lesions associated with tertiary syphilis. In 1494, King Charles VIII of France invaded Italy with Spaniards and other mercenaries. At first there was little fighting and much cavorting with camp followers, resulting in an outbreak of syphilis in Naples that decimated the army (including King Charles) and then spread throughout Europe. Because this initial surge coincided with Columbus's return from the New World, historians have argued ever since about whether *T. pallidum* hitched a ride back inside a few randy sailors. For about fifty years it raged with epidemic intensity, affecting as much as 20 percent of urban populations, but then seemed to lose strength and settled into a chronic, degenerative form with fewer fatalities, though still hideous. "A man can no more separate age and covetousness than part young limbs and lechery," Shakespeare wrote in *Henry IV,* "but the gout galls the one, and the pox pinches the other." Its pinched victims included Pope Alexander Borgia, Henry VIII, Peter the Great, Benvenuto Cellini ("I believe I caught it from that fine young servant-girl whom I was keeping," he recalled in his autobiography, observing that it covered his body with rosy blisters "the size of six-pence"), Toulouse-Lautrec, Baudelaire, Guy de Maupassant, Nietzsche, and Heinrich Heine. Ivan the Terrible's reputation was sealed by the madness of late-stage syphilis. "Everybody has it, more or less," quipped Flaubert.

By the seventeenth century it was being confused with a lesser demon, gonorrhea. It even gained a reputation for imparting brief periods of feverish creativity, as Thomas Mann wrote in *Doctor Faustus.* For years, mercury was considered the best remedy for syphilis, leading to innumerable cases of poisoning that were as awful as the disease (hence the saying "One night with Venus, a lifetime with Mercury"). For a while, sarsaparilla was thought to be effective, perhaps explaining why otherwise tough cowboys asked for it at the saloon. *T. pallidum* was not identified until 1905, after which various drug therapies radically reduced its incidence in the developed world. The German bacteriologist and Nobel Prize winner Paul Ehrlich (1854–1915), searching for an all-purpose bacteria killer, coined the term *Wünder-Bullet* ("magic bullet") to describe his 606th attempt, an arsenic compound known as Salvarsan, which was effective against *T. pallidum* but produced bad side effects. He persevered to create a safer drug, "914" or Neosalvarsan, which helped combat syphilis during World War I. Still, during the 1920s more than nine thousand Americans died of syphilis and sixty thousand babies were born with *T. pallidum* every year. It finally brought down Al Capone.

Like other sexually transmitted diseases, syphilis tends to accompany promiscuity, drug abuse, and poverty. In a ghastly episode that illustrates how groups can become stigmatized by such afflictions, several hundred African-American men suffering from syphilis in Georgia in 1932 were allowed to develop advanced stages of the disease over a period of decades for the sake of supposed research under the auspices of the U.S. Public Health Service. Known as the "Tuskegee experiment," it had little or no scientific value, given that symptoms were already well documented. Penicillin that could have halted the disease was available, of course, but was withheld. Nazi Germany surpassed such medical scandal only in numbers.

By 1970, syphilis was killing only two of every million Americans. But the number of cases had begun to climb in the late 1960s, and crack cocaine fueled an alarming increase in the late 1980s. In 1997, federal health officials announced a national plan to eliminate the disease, defined as fewer than a thousand cases by 2005. By 2000, the overall count of sexually transmitted syphilis was below six thousand, but still rising among gay and bisexual men, especially in big cities. By 2007, 11,466 primary-stage (first symptoms) and secondary-stage (one to six months later) cases were reported. The World Health Organization estimates that twenty-eight million cases still exist worldwide, with twelve million new sexually acquired infections every year. Condoms are still the second most certain way to avoid syphilis.

One of *T. pallidum*'s close relatives (so close as to be almost identical), *T. pertenue,* causes yaws, a nonvenereal raspberry-like rash encountered mainly in equatorial countries. Unlike syphilis, there is ancient evidence of yaws in both the Old and New worlds, leading some historians to conclude that yaws was always a village disease, while syphilis required the social conditions of urbanization. It was so common among African slaves in America that "yaw houses" were built to isolate them. In the 1950s and '60s, penicillin almost eradicated yaws, but it has come back, especially in West Africa. Millions of people, mostly children, catch it every year.

Another family member, *T. microdentium,* is a harmless inhabitant of our mouths.

• **tinea** •

Called ring*worm,* tinea is actually a fungus. Such knowledge provides scant comfort.

Tinea capitis, the scalp variety, was epidemic in North America during and after World War II. Grungy movie theater chairs were often responsible for starting an outbreak, when lots of little boys with wiffle haircuts would scrunch down in their seats every Saturday, either picking up or leaving *Microsporum* or *Trichophyton* fungi (long-haired girls were less likely to get infected). In severe, big city circumstances, doctors prescribed X-ray treatments to kill the fungus. "The rays must make infected hair fall out," intoned the always gung-ho *Time* magazine in December 1943, not seeing the cancer lawsuits on the horizon. Children in Sault Ste. Marie, Ontario, had to wear white cotton bonnets when ringworm took up residence on the heads of 1,363 of the city's 5,712 elementary school children in 1950. The bonnets were boiled every night by horrified moms. Happily, various ointments and salves have largely chased away the itchy, flaky ringworm-of-the-scalp blues since then, though it can still occur where hygiene is poor. It can also be picked up from pet cats and dogs. In 1992, NCAA wrestlers had to submit to scalp exams before they could compete in championship matches.

Tinea pedis, or athlete's foot, is another well-known *Trichophyton* infection, familiar to millions of television viewers who will never see a real case. Tinea cruris, called "jock itch" by cognoscenti, tends to bother men whose clothes are too tight or who are just plain old fat. Tinea barbae is an unusual infection of beards, often bacterial but sometimes fungal. There are several more kinds of tinea, but isn't this enough for now?

• *Trichinella* •

The intestinal roundworm *Trichinella spiralis* can get into human muscle tissue if we eat uncooked, infected meat, a condition known as trichinosis. Pork is the classic bearer of bad news, though bear and walrus meat are on record as hurting a few adventurous souls. Over a period of four to six weeks after initial exposure, abdominal symptoms like nausea and diarrhea may occur, as the microscopic creatures go through a life cycle that eventually puts millimeter-long larvae in tongue, eye, pectoral, and other muscles, causing pain, fever, hemorrhaging, and occasionally death. Trichinosis is easily prevented by cooking meat thoroughly. It is uncommon in the United States and almost unheard of in countries such as France, where pigs dine on root vegetables instead of garbage. Laws prohibiting the feeding of raw meat scraps to hogs have helped lower the number of cases reported in the United States, where wild game is responsible for more infections than pork. The biggest contemporary American outbreak, involving ninety cases, took place in 1990 when Southeast Asian immigrants ate raw pork sausage at a wedding reception.

• tuberculosis •

Tuberculosis, or TB, is one of the great common denominators of human life—one-third of the world's population is infected—yet is paradoxically one of the most easily combated. In the last two centuries it has killed about a billion people, but never with the obvious force of social upheaval that characterizes other great epidemics like smallpox or cholera. The World Health Organization estimates that there were 9.27 million new cases of TB and nearly 2 million deaths globally in 2007, with 13 of the

15 highest-incidence-rate countries found in Africa—a tragedy linked to high rates of HIV coinfection, which reaches 73 percent in South Africa. Three species of mycobacteria can cause it—*Mycobacterium tuberculosis, M. bovis,* and *M. africanum*—and they have probably been at work since the dawn of humankind.

Evidence of "phthisis," as Hippocrates called it, or "consumption" has been found in neolithic (4500 B.C.) skeletons in Germany and has been written about in diverse ancient texts (except those of the Americas), including the Hammurabi Code of circa 2000 B.C. Historians believe the disease evolved in the Middle East about eight thousand years ago and crossed over to humans from cattle. It may have reached the Americas via migration from Asia across the Bering Sea. Mummified remains of a woman who died a thousand years ago in Peru show typical lung scars.

For hundreds of years in England, scrofula—now known to be a mycobacterial infection of the glands—was called "the king's evil" and was treated by divine "royal touch" from the monarch himself until the eighteenth century. But tuberculosis (the world "tubercle" means a small protuberance, nodule, or growth) did not become a major problem until the Industrial Revolution created the squalid urban conditions it needs to spread. Already in the late seventeenth century, TB was killing one out of every five Londoners, according to the philosopher John Locke. The disease ravaged Native Americans with added intensity after exposure to Europeans and the forced change from nomadic to reservation living. René-Théophile-Hyacinthe Laënnec (1781–1826), the inventor of the stethoscope—his creation was a hollow wooden tube, not the binaural instrument familiar today—and one of many prodigious tuberculosis researchers in France, died of the disease. In 1900, TB was killing 200 of every 100,000 (mostly urban) Americans, but as public health measures improved, the rate fell to 60 per 100,000 by 1940 on the eve of antibiotic therapy. A total of 12,904 cases (4.2 per 100,000) were reported in 2008.

Coughing, sneezing, talking, singing—anything that makes the bacteria fly out of a carrier's lungs or throat can bring exposure, but infection most often requires months of close contact. Animals also get and pass TB—humans can pick it up from drinking infected cow's milk, though this is rarely a problem where pasteurization is the rule.

The slow-growing tuberculosis bacillus was identified in 1882 by the great German bacteriologist Robert Koch, who also found the organisms responsible for anthrax and cholera. The object of worshipful veneration in America, Europe, and Japan

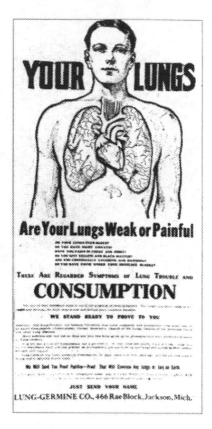

(where a shrine was built for those who truly believed in his godlike status), Koch took a professional tumble when he later announced a TB vaccine that turned out to be a dangerous dud. Nevertheless, his "tuberculin" did form the basis for today's simple skin-prick test for tuberculosis—in itself a major development. This test plus chest X rays became the primary tools for tracking the disease.

Until the late 1940s, when the drug streptomycin was found to control it, most sufferers were simply told to rest outside where the air was clean, often at rural hospices designed for that purpose. Galen, the legendary Greek physician of Rome (A.D. 130–

200), sent his rich consumptive patients to the beaches of Stabia, across from the Isle of Capri. Thomas Mann wrote about one of the illustrious German sanatoriums in *The Magic Mountain*. TB's victims could constitute a pantheon of the high arts (Keats, Shelley, Chekhov, Chopin, Emily Brontë, Lawrence, Orwell, Stevenson, Whitman), thereby contributing to the romance of creative genius and suffering. "Decay and disease are often beautiful, like the pearly tear of the shellfish and the hectic glow of consumption," Thoreau wrote in 1852. This notion was epitomized by the consumptive courtesan Violetta Valéry in Verdi's 1853 opera, *La Traviata,* based on Alexandre Dumas's novel *La Dame aux Camélias.*

But today more than ever TB is an affliction primarily of the urban poor who do not have access to effective drug regimens. *M. tuberculosis* is a supreme opportunist, waiting for any glitch in the immune system to give it a chance to strike. Malnutrition is often the door opener, along with chronic infections by other organisms. In the 1980s, the number of new cases in the United States rose steadily to about 26,000 a year, due to a breakdown of treatment services, increasing numbers of infected immigrants, multidrug-resistant strains, and the emergence of AIDS. In 2008, most U.S. cases occurred among the foreign-born population (especially those from Mexico, India, China, Vietnam, and the Philippines), and Asian Americans had by far the highest case rates. Among native-born Americans with TB in 2007, 45 percent were non-Hispanic black or African-American.

Compared to some other germs, such as those that cause measles or mumps (where one carrier can infect eighty out of a hundred susceptible people), TB bacteria are rather weak. The disease is thus not highly communicable under decent living conditions—just letting fresh air blow through a room will remove almost two-thirds of the infectious material exhaled by a sick person every day. Because the bacteria are sensitive to ultraviolet rays, infection rarely happens outdoors in daylight. Only half of the people who cohabit with someone with infectious TB will likely catch it. But overcrowded American jails are TB hothouses. Prevention is complicated by the fact that at least twelve weeks must pass before a person exposed to the disease will test positive. Moreover, people who test positive may be passive carriers of the disease for years until they develop active TB and thus become infectious to others. A vaccine known as BCG (for Bacillus Calmette-Guérin), named after the two French scientists who developed it, is not often used in the United States but may be given to children in countries where tuberculosis is common.

In 1992, an airline flight attendant gave TB to two crew

members out of 274 with whom she had worked for varying periods, raising questions about air quality in the confined space of airplane cabins. The two infected persons had spent only about twelve hours with her. In 1994, four travelers on an eight-and-a-half-hour flight from Chicago to Honolulu caught TB by sitting in the same section as an infectious passenger, leading the Centers for Disease Control and Prevention to advise that people with active TB should not be allowed on commercial flights.

Since treatment is long and expensive, requiring daily drug doses for six to twelve months in conventional cases, shoddy care often results in microbial resistance to the frontline medicines used to fight TB, a serious problem today in the United States. In October 1994, the federal government issued new guidelines for prisons, hospitals, nursing homes, and medical and dental offices to curb the spread of drug-resistant strains. The standards require better ventilation, more efficient masks for workers, and use of ultraviolet light to kill germs. All in all, tuberculosis is an accurate measure of how a society has been doing in the contest between greed and compassion.

Mycobacterium leprae, a cousin of the TB germs, causes leprosy, found most often in Asia and Africa. Since the 1970s, the illness has also been known as Hansen's disease, after Armauer Hansen (1841–1912), the Norwegian bacteriologist who first diagnosed it in detail in 1873. Though the Bible mentions a disfiguring disease assumed to be leprosy, archaeological studies have found no skeletal signs of it before A.D. 500. Leprosy was rampant in Europe from around 1200 until the Black Death era in the fourteenth century, when the sheer loss of human hosts probably stymied *M. leprae* and perhaps opened a niche for *M. tuberculosis.* Today, about 1 to 2 million people around the world are permanently disfigured by leprosy. The World Health Organization tallied 212,802 cases in 2008, with stubborn pockets of endemic infection in Brazil, India, Nepal, and several African countries. In 2002, 96 cases were reported in the United States.

M. leprae has never been cultured in the usual laboratory media, but it will grow in the foot pads of mice. It may reside in armadillos. Nature's above art.

• typhoid •
(Salmonella typhi)

The word "typhoid" means "stuporlike," which describes the mental state of untreated victims. It is caused by a bacterium, *Salmonella typhi,* that comes in more than fifty strains and inhabits only the human digestive system. Though the disease is uncommon now in the United States, its name is firmly entrenched in the lingua franca because of one Mary Malone, a cook who never herself fell ill with the disease but spread it far and wide and became known as Typhoid Mary.

Typhoid is a filth disease usually caught from contaminated food and water. Thus, like typhus, with which it is often confused, typhoid has always tended to appear when hygiene and sanitation break down due to war or poverty (though a Stanford University frat house experienced an outbreak in the late 1960s when a toilet leaked into the kitchen). During the Boer War, *S. typhi* killed 13,000 British soldiers (compared to 8,000 lost in combat) and sent another 64,000 home early. In the Spanish-American War, 20 percent of U.S. forces got typhoid fever, with more than 1,500 deaths. By World War I, doctors understood the role of infected feces in the spread of the disease, and supplies of fresh water and toilet paper to the troops led to a sharp drop in cases. During this era, typhoid was common in American towns and cities, with especially high death rates in Pittsburgh, Pennsylvania, and Troy, New York.

Because of Typhoid Mary, *S. typhi* is infamous for its ability to reside for long periods of time without producing illness in its host while being able to infect other people and cause typical symptoms. This obviously complicates any effort to control the disease. Mary Malone came to America from Ireland as a teenager and was employed as a cook in well-to-do households In the winter of 1906–7, an epidemiologist and sanitary engineer named George Soper was hired by the owner of a home in Oyster Bay, New York, to investigate a typhoid outbreak there in a wealthy family that had rented the house during the previous summer. Soper tested the obvious water sources but found no pathogens. He finally interviewed Malone and discovered through her work record that her employment had coincided with seven separate outbreaks of typhoid. When he confronted her with this evidence, she threatened him with a carving fork. She was, after all, in excellent health. Soper then recommended to New York City health department officials that Malone be arrested. This was accomplished after a more violent fork scene with city police, whereupon she was isolated for testing and treatment.

Hearst newspapers soon sensationalized the story of "Typhoid Mary," a sobriquet created by city health officials. After spending two years in a hospital bungalow on North Brother Island, she was released under legal pressure in 1910 on a promise not to resume her trade as a family cook. She promptly adopted aliases and found regular work in hotels and restaurants. Her infective trail ended ironically in 1915 at the Sloane Hospital for Women, where 20 cases of typhoid were blamed on her kitchen job. She was confined again on North Brother Island until 1932, when a severe stroke sent her to a Bronx hospital, where she died of pneumonia six years later. In all, she was directly linked to at least 53 cases of typhoid, three of them fatal.

Today, some 400 cases of typhoid are reported every year in the United States, mostly among people who have traveled in Third World countries, where it strikes about 22 million people a year, with 216,000 deaths. There are believed to be about 2,000 chronic carriers in this country, most of them elderly women, some like Mary Malone with no history of clinical ill-ness. Typhoid can have a mortality rate between 12 and 30 per-cent without treatment, but antibiotics bring it down to fewer than 1 percent of cases, usually among the very young or old and the malnourished. Besides vaccination that provides several years of protection, clean water, pasteurized milk, and sanitary food handling are the best prevention. The disease is still com-mon in Africa, Southeast Asia, Central and South America, the Middle East, and India—in other words, wherever masses of people live together with little or no sanitation.

• typhus •

"Typhus," from the Greek word *typhos,* meaning "the stupor caused by fever," refers to two types of fever-rash illness. Epidemic typhus, also known as jail fever, is caused by *Rickettsia prowazeki* microbes carried by body lice from person to person. Murine (or endemic) typhus is caused by *Rickettsia typhi* brought to humans from rats by the same flea that can transport bubonic plague. They are both associated with filthy, crowded living conditions.

For centuries, epidemic typhus was a loyal follower of war and famine. The body louse (*Pediculus humanus,* aka "cootie") feeds on infected blood, then jumps to other hosts, where it defecates live *R. prowazeki* organisms that get scratched into the skin. Stop-ping the spread of disease entails getting rid of the lice. None of this was understood until the first two decades of the twentieth century, when the researchers Howard T. Ricketts and Stanislas von Prowazek did pioneering work. Both died of typhus.

The first record of typhus was an outbreak in a Sicilian monastery in 1083. Throughout the Middle Ages, it raged across Europe wherever war and revolution fostered the squalid condi-tions typhus requires. Explorers and traders took it to the New

World, where it helped exterminate native tribes. Columbus caught typhus in 1494. It added to the immeasurable catastrophe of the Thirty Years' War in Germany in the seventeenth century, gutted Napoleon's army and its opponents in Russia (30,000 ill French were left to die in Vilna), and arrived in East Coast American cities as "shipboard fever" among immigrants. In 1850, Zachary Taylor is thought to have died of typhus after just sixteen months in the presidency (some historians now believe he might have had Rocky Mountain spotted fever, also caused by a rickettsia germ). Typhus took hundreds of thousands of victims across Eastern Europe during World War I. History's worst typhus epidemic occurred in Russia around the time of the Bolshevik revolution—between 1917 and 1921, there may have been 25 million cases, with 3 million deaths. The eighteenth-century Russian soldier and politician Grigori Potemkin died of typhus.

During and after World War II, typhus returned to strike Italy, Japan, and Korea, but the most wretched scenario unfolded in Nazi concentration camps. When Bergen-Belsen was liberated in the spring of 1945, nearly all of the 55,000 inmates had suffered from typhus over the preceding winter or were currently afflicted, resulting in thousands of deaths (mortality can be as high as 60 percent among older untreated patients). Delousing, which consisted of dusting with the insecticide DDT, quickly brought the outbreak to an end.

Today, antibiotics easily take care of typhus if used early on. Vaccines have been developed, but none is commercially available in the United States and no country requires them for entry. Epidemic typhus is rare except when there are long breakdowns in public hygiene, while endemic typhus occurs year-round in the tropics. Dusting with pesticides to get rid of lice is still practiced in highly infested areas, but being powdered with potent chemicals like DDT, lindane, or malathion is nothing to look forward to except under extreme duress.

The ability of full-blown typhus to make victims disoriented and deranged found its way into the modern Italian language: *tifo* means "typhus," and the verb *tifare* means "to cheer" or "to be a fan." Thus the slogan of the Forza Roma soccer team is *Tifare è un dovere*—literally, to be a fan is a duty.

• varicella–zoster virus •

Chicken pox, caused by the varicella-zoster virus, is the last holdout of the common childhood diseases. About 90 percent of Americans have suffered through an itchy, grumpy week of it by the time they reach adulthood. For the vast majority it is a largely benign experience, with the greatest social cost calculated from how much work time parents lose while caring for scabby kids at home.

The varicella-zoster virus, a member of the herpes family, persists for decades in the body after oatmeal baths are just a yucky memory. Contrary to conventional wisdom, the ubiquitous childhood attack does not give permanent immunity. Later bouts are usually so mild as to be without noticeable symptoms. Past the age of fifty, however, as the immune system weakens, the latent virus is more and more likely to trigger a painful skin rash called herpes zoster, or shingles (from the Latin word for belt, *cingulum,* that describes the rash's characteristic band around the ribs). Some 300,000 cases occur yearly in the United States. A study of age incidence has shown that half of people who live to be eighty-five will experience this *recherche du temps perdue,* which is about 25 percent as infectious as chicken pox. Yes, a grandparent with shingles can give chicken pox to previously unexposed kids.

Chicken pox was not recognized as being different from smallpox until the late nineteenth century. In the 1970s, Japanese scientists developed a chicken pox vaccine; a similar shot

Faithful companion of childhood, the chicken pox virus.

was licensed in the United States in 1995 and, with vaccination coverage among children aged nineteen to thirty-five months rising past 60 percent, cases nationwide have declined dramatically.

• West Nile virus •

In June 1999, the crows started dying in Queens. By the end of November, New York State alone tallied 13,654 reports of 17,339 dead birds from thirty-two counties. Abnormally high numbers of wild birds also plotzed in Connecticut and New Jersey. The kill was especially disastrous for crows, but other species were hit, too, including blue jays, hawks, mallards, and four Chilean flamingos at the Bronx Zoo. Horses on Long Island came down with a neurologic illness. And beginning in late August, humans started to get sick with what was first diagnosed as St. Louis encephalitis, but soon confirmed as West Nile fever, previously unknown in the Western Hemisphere. Sixty-two cases of severe illness and seven deaths occurred in New York by the end of the year.

The virus behind this mayhem was first isolated from a woman in the West Nile region of Uganda in 1937. A flavivirus transmitted to humans and animals by mosquitoes that pick it up from biting birds, it has been endemic in Africa, Europe, the Middle East, and Asia for years. In the United States, *Culex* mosquitoes are mostly to blame, though other kinds may also carry the West Nile virus. It has been found in more than seventy species of birds. Genomic tests indicate that it arrived in New York from Israel, where deaths of domestic geese and two people occurred at the same time as the American outbreak. It quickly spread along the East Coast, and the assumption is that it is now firmly established in the New World.

Fortunately, most West Nile infections are mild, with symptoms resembling flu. The most dangerous possibility is encephalitis. Even in areas where the virus is endemic, fewer than 1 percent of mosquitoes are carriers, and fewer than 1 percent of people who are bitten and become infected will get severely ill. Death rates in these cases range as high as 15 percent, primarily among elderly victims.

As with other insect-borne diseases, the key to limiting the disease is controlling the insects. West Nile is not contagious between people, between people and susceptible animals (horses, cats, bats, squirrels, rabbits, and a few others), or between animals. No mosquito bite, no West Nile fever.

• worms •

Various nematodes, trematodes, and cestodes—worms to you, bub—cause enormous suffering in poor countries around the world. A combination of climate, unsanitary living conditions, and ignorance puts billions of people at risk for debilitating and potentially lethal disease from microscopic worms.

Guinea worm disease, caused by the nematode worm *Dracunculus medinensis,* comes from contaminated drinking water. Well over a hundred million people are at risk for it in India, Pakistan, and many African countries. At least two million cases exist. A year after a victim drinks invisible parasites left by water fleas, adult worms over *one yard long* crawl out through the victim's skin in excruciating, science-fictional horror fashion. The infection itself is not fatal, but bacterial complications can be.

Lymphatic filariasis, caused by the parasitic worms *Wuchereria bancrofti* or *Brugia malayi* in the human lymphatic system and carried in mosquitoes, infects some 120 million people in tropical and subtropical regions of the world. The World Health Organization ranks it as humankind's second most common cause of disability. Again, the infection is not fatal, but the presence of adult worms in the lymph system for as long as ten to fifteen years can produce elephantiasis.

Onchocerca volvulus nematodes cause another kind of filariasis called river blindness, transmitted by blackflies. Eighteen million people in Africa, Yemen, and Latin America have the worms under their skin, which has led to blindness in some 270,000 cases.

Taenia saginata (beef tapeworm) and *Taenia solium* (pork tapeworm) infect humans through undercooked meat. Adult *T. saginata* can live for more than thirty years in human intestines and grow to seventy-five feet long. Fifty million people in Latin America, Asia, and Africa carry both parasites, resulting in 50,000 deaths annually as larva create cysts in muscle and other tissue.

Ascaris lumbricoides, an intestinal worm that causes the disease called roundworm, is staggeringly common in the human race, infecting a billion people in tropical and subtropical countries and killing 20,000 a year through intestinal obstruction. Soil and water contaminated by human feces spread the parasite.

Ancylostoma duodenale and *Necatur americanus* are two of many parasitic worms that cause hookworm disease. *A. duodenale* infects an estimated billion people in the tropics, where 60,000 die annually. *N. americanus* used to be common in the southeastern United States. Picked up through direct contact with larvae in contaminated soil, often by walking barefoot, the worms multiply in the intestines and eventually eat so much blood that the victim gets anemic. Human feces is again the transmitting material.

Parasites in the genus *Schistosoma* infect about 200 million in Africa, Asia, and Latin America through skin contact with water contaminated by human feces or urine. Ever-popular dam and irrigation projects, usually financed by rich countries, have helped spread the problem by widening the habitat of snails that are part of the worms' life cycle.

Finally, *Enterobius vermicularis,* the pinworm, infects countless

Ascaris lumbricoides, *or roundworm, which can grow to be more than twenty centimeters long, infects a billion people worldwide.*

humans, especially children, in temperate as well as tropical areas. It is the most common worm infection in the United States. The parasites live in the rectum for at least three months, then lay eggs near the anus, which prompts itching and scratching. (Hence the simple "scotch tape test" diagnosis, whereby eggs are captured on a piece of said transparent tape stuck over said orifice.) Fingers then go to mouths, where the cycle continues. They can cause appendicitis and female genital tract problems.

There are more, but these are the big ones. Some might be eradicated with sufficient money and political willpower; others will never go away. They are a constant reminder of worldwide poverty.

• XMRV •

The *x*enotropic *m*urine leukemia virus-*r*elated *v*irus is a member of the dreaded retrovirus family, in the genus *gammaretrovirus* along with the devil that causes feline leukemia. It was discovered by researchers from the University of California at San Francisco and the Cleveland Clinic in 2006 in the prostate tumors of patients with a specific genetic risk factor for prostate cancer. Gammaretroviruses had been studied in animals for many years, but this was the first one found to infect humans.

XMRV has since been noticed in about a quarter of biopsies done on men with prostate cancer (the most common non-skin cancer in American males), compared with only 4 percent of benign biopsies. It seems to be particularly associated with highly aggressive tumors, but no firm causative link has yet been proven. In 2009, XMRV was also located in 67 percent of patients with chronic fatigue syndrome compared with just 3.7 percent of healthy subjects. As with prostate cancer, more research will be needed to establish whether the virus is to blame for the disease. But with 3 to 4 percent of healthy control groups in both areas of study having been found to carry XMRV, this could translate into some 10 million infected Americans—clearly a cause for concern. Like the HIV retrovirus, XMRV is thought to be transmitted via bodily fluids but not through the air.

• yellow fever •

Yellow fever was the American plague, one of the great tropical fevers that shaped the development of the New World after Columbus. Whether his crew introduced it or not, the yellow fever virus helped exterminate existing civilizations, thus facilitating European conquest, even though whites were as vulnerable as aboriginals. Disastrous urban epidemics in Philadelphia, New Orleans, Charleston, and other blossoming coastal cities tempered nativist optimism about the continent's natural promise. Other diseases may have taken more lives, but none had more cultural or political impact. "Yellow fever will discourage

An Aedes aegypti *mosquito (not actual size), first identified as the carrier of yellow fever in 1881.*

the growth of great cities in our nation," Thomas Jefferson worried in 1800. In fact, it helped drive home the need for basic public services like clean water supplies, sewage systems, and health departments.

The virus is transmitted by certain female *Aedes* mosquitoes, which were not native to the Americas. *A. aegypti,* the classic carrier, likes to lay its eggs in containers of clean water, so it thrives around human settlements, especially in humid climates above 72°F. Epidemics in Barbados in 1647 and in the Yucatan and Cuba in 1648 signaled that both microbe and insect had made the slow voyage from West Africa and established themselves in the Western Hemisphere (the relative immunity of black slaves in the Caribbean suggests yellow fever's origin in Africa). In 1693, Cotton Mather described characteristic symptoms—yellow skin, black vomit—of a disease introduced to Boston by a British vessel from Barbados. Previously unexposed individuals had a fifty-fifty chance of survival. Charleston and Philadelphia were struck for the first time in 1699. New York lost 10 percent of its population in 1702. Transatlantic trade guaranteed that yellow fever would bounce back toward Europe, where the first outbreak occurred in Cádiz, Spain, in 1730 upon the return of a fleet from Cartegena, Colombia: 2,200 people died.

Philadelphia suffered a paroxysmal outbreak of yellow fever in 1793 that altered the young nation's collective consciousness. Fifteen percent of the city's population succumbed, with more than 4,000 lives lost between August and the arrival of frosty weather in October. People who did not catch the disease chose to flee, virtually emptying the city and prompting New Jersey, Delaware, and Maryland to try to keep out Philadelphians. As during the Black Death of the Middle Ages, the sinews of society were wrenched apart, with family members abandoning each other, officials ordering counterproductive measures, natives blaming foreigners (in this case German and Haitian immigrants), clerics finding reasons not to assist the helpless, and corrupt merchants exploiting everyone. A provisional government was established to maintain some city services in the absence of most officeholders. Dr. Benjamin Rush, the country's most eminent physician and a powerful proponent of the "miasma"—that is, environmental—theory of disease, thought for a while that yellow fever stemmed from a load of rotten coffee beans dumped on the docks. There was no mass violence, however. Other East Coast cities also experienced serious yellow fever epidemics, and the disease no doubt hastened Philadelphia's slippage behind New York as an important trading port.

Yellow fever had momentous consequences elsewhere in the world, as well. In 1802, Napoleon tried to defeat a slave revolt in Haiti with 25,000 troops but lost 22,000 of them to yellow fever. The debacle soured Napoleon on the New World, contributing to his sale of the Louisiana Territory in 1803, thus doubling the area of the United States. With no army of occupation, Haiti declared independence in 1804, becoming the first black nation to free itself from European colonial rule.

In Africa, the yellow fever virus was a keystone in the microbial barrier that prevented European exploration of the continent until the latter part of the nineteenth century. African slaves were known to be relatively immune to yellow fever, malaria, and other diseases that ravaged Europeans and native Americans, though the reasons were of course totally obscure. White surgeons aboard "Guineaman" slave ships had observed these afflictions along the African coast without being able to discern different symptoms. Yellow fever was so common on trade ships between Africa and America that it was dubbed "Yellow Jack" after the flag flown from quarantined vessels. Native Africans incorporated into folk music the role of fever and mosquitoes in protecting them from white colonists.

Until the link between yellow fever and *Aedes aegypti* was proved in 1902, there was no reliable control of the disease. Many observers had drawn the right conclusion empirically; Dr. Carlos

A nativist nineteenth-century allegorical cartoon showing Columbia rescuing Florida from Yellow Jack, who has apparently sprung from the crate labeled Trade.

Finlay, a Cuban physician, even zeroed in on *Aedes* as early as 1818. The work of Dr. Walter Reed's Yellow Fever Commission, supported by recent proof of mosquito involvement with malaria, finally removed all doubt. Commission members and military volunteers who allowed themselves to be bitten by infected mosquitos were the heroes of the research. The disease was soon halted in Cuba, Rio de Janeiro, Panama, and Brazil. Ironically, a campaign for worldwide eradication, spearheaded by the Rockefeller Foundation in 1913, met defeat in yellow fever's primordial breeding ground. In Africa and South America, a jungle variety of the disease can pass between monkeys and humans via a number of forest canopy mosquitoes. Breaking this circle of infection was out of the question. A vaccine was developed by 1937 and given on a mass basis beginning in French West Africa in 1939. An epidemic in Ethiopia between 1960 and 1962 infected more than 100,000 people, killing a third.

Today, yellow fever occurs mostly in underdeveloped, rural areas where control measures and/or immunization are lax. It thrives only in Africa—an outbreak in Kenya in 1993 underscored its continuing threat—and South America, where it probably causes more than 30,000 deaths every year. Mortality can be as high as 10 percent in large outbreaks, though the disease is often mild enough to escape detection.

• *Yersinia pestis* •
(plague)

The bacterium *Yersinia pestis* was once the planet's most nightmarish life form, making all lions, tigers, and bears seem like kid's stuff. The capital-P Plague, aka Black Death, caused by this bacterium set the physical, social, and metaphorical standards for our perception of epidemic disease. Today it rarely occurs (in the United States, mostly in Arizona, New Mexico, and Col-

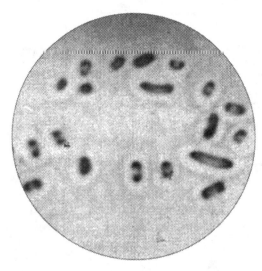

Plague bacteria, once capable of destroying civilization, magnified a thousand times.

orado) and can be controlled by antibiotics when it does. But in its heyday, *Y. pestis* more than earned a place among the Four Horsemen of the Apocalypse.

Bubonic plague comes from the bites of various types of fleas (principally *Xenopsylla cheopsis,* the Oriental rat flea) that have fed on infected animals. ("Bubonic" refers to the *buboes,* or incredibly painful swollen lymph nodes, that are an early symptom.) The animals include squirrels, marmots, prairie dogs, rabbits, mice, house cats, and, classically, the common domestic rat, *Rattus rattus.* No rats, no fleas, no plague. This relationship, which sounds so straightforward, was not fully explained until 1914. For at least fifteen hundred years, people applied every imaginable combination of folk wisdom, magic, and guesswork to combat the disease. As has often been the case in medicine, the grotesquerie of the "cures" only reflected the desperation of the sick. Drinking witches' brews of snakes and gemstones, inhaling the stench of human feces, building houses to face

Devoted scientists, examining rats for bubonic plague in New Orleans, 1914.

away from ill winds—when nothing works, anything is worth a try. The nursery rhyme "ring-a-ring o'roses" (or "ring around the rosey") may hark back to the belief that a fragrant bouquet of flowers and herbs would ward off the stinking plague. A few public health innovations, like street cleaning and quarantines (from the Italian: In 1348, Venice ordered all travelers from plague-beset regions to be isolated for forty days, or *quarantena*), were somewhat helpful, but many just made matters worse.

Y. pestis probably moved into its mobile rodent home in primordial Africa or India. Homer's *Iliad* mentions a disease linked with rats during the Trojan War (1190 B.C.) that might have been plague. But the first reliable report was Procopius's description of the so-called plague of Justinian in A.D. 542–43, which arrived in the Mediterranean from Egypt or Ethiopia. Deaths were said to reach 10,000 a day in Constantinople, with

"The Plague Pit," an early-nineteenth-century engraving, depicts victims being dumped from a cart into a mass grave after dark.

20 to 25 percent of the region's population wiped out over a five-year period. Some reports claimed that half of the Roman Empire's subjects died, hastening its fall. This was the first of three mighty waves that left clear evidence as *Y. pestis* moved around the planet with commerce, war, and the bacteria's own wicked pulse.

The second, called the Black Death after the gangrenous extremities of its victims, began in the fourteenth century. It started on the Mongolian steppes and spread westward with commerce across Asia, having already reduced the population of China from 123 million in 1200 to 65 million in 1393. It may ultimately have killed a third of all Europeans (20 million to 30 million people). The infamous story of plague-infested corpses being catapulted over the walls of besieged cities by encircling armies stems from the Tatar attack on Kaffa (now Feodosiya, Russia) in the southeast Crimea in 1346. Carried aboard ships by rats and fleas, *Y. pestis* spread around the Mediterranean and through-

out Europe during an epoch when medieval political and eco-
nomic systems were already strained by population growth and
other social problems, thus helping quite literally to clear the way
for new ideas. Some of the seminal cities of the budding Renais-
sance—Florence, Venice, Siena, Orvieto—lost perhaps 50 to 60
percent of their inhabitants. Boccaccio wrote (not as an eyewit-
ness, however) of the destruction of traditional authority at all
levels, from family to state. Paris had a death rate of 800 a day. Jews
were exterminated as scapegoats to an extent not seen again until
the Nazi Holocaust—in Strasbourg alone, 16,000 were murdered. As late as 1656, Naples lost 300,000 citizens to the disease
in five months. In August and September 1665, London lost
7,000 a week to the great plague. "Lord! How sad a sight it is to
see the streets empty of people," Samuel Pepys wrote on August
16 of that year. The Great Fire of 1666 helped to end the carnage
by destroying rat habitats in thatched roofs. Moscow suffered the
final slaughter by *Y. pestis* in Europe in 1771.

The third epidemic started in China during the mid–
nineteenth century. Research teams ventured there in the early
1890s, when Swiss-French bacteriologist Alexandre Yersin (1863–
1943) discovered the responsible organism in Hong Kong. The
germ made its way to the western United States probably around
1900, caused an epidemic in Los Angeles in 1924–25, and ran on
through World War II. The insecticide DDT largely stopped its
spread, yet it endures in ground squirrels and other wild rodent
burrows, infecting ten to fifteen people a year. Since the last giant
death toll of greater than 10 million during the first decade of
the twentieth century in India, outbreaks there have been relatively limited. During the anarchy that followed Indian independence in 1947, 57,000 people died in a single state. Thousands
of cases of plague were reported during the war in Vietnam.
Today, about 1,000 to 3,000 cases are reported by the World
Health Organization every year.

Dream on—an angel tells Charlemagne of a plant that can stop the Plague.

Although another epidemic of Black Death proportions is extremely unlikely, multiple cases of plague could still appear wherever people live in crowded, unsanitary circumstances near infected rodents. Helter-skelter suburbanization in the Southwest has been a prime factor behind cases in the United States. An infected rat and two infected fox squirrels (*Sciurus niger*) were found in Dallas in 1993. The mortality rate for untreated bubonic plague is about 60 percent. The single American fatality in 1993 did not receive medical attention until six days after the victim started feeling ill, at which point he died within six hours of entering the hospital.

Pneumonic plague results when *Y. pestis* enters the lungs via the breath of someone else already infected instead of via a flea bite. Death is virtually certain if treatment is not begun within twenty-four hours, despite the organism's vulnerability to many common antibiotics (except penicillin). In September 1994, more than 400,000 residents of the western Indian industrial city of Surat fled an outbreak of pneumonic plague believed to be the first anywhere in the world for decades. A symbol of squalid commercialization known for its pervasive stench of refuse, excrement, animal carcasses, and textile-processing chemicals, Surat produced a thousand tons of garbage every day but disposed of less than half of it. Just one-eighth of the city's area had sewage service. Vans roamed panic-stricken neighborhoods urging the entire population of 2 million to stay put and take free handouts of the antibiotic tetracycline, which quickly became a black market commodity. Riot police were called upon to enforce an 1897 plague control law allowing forced hospitalization of victims and control of crowds around hospitals and pharmacies. As in days of old, doctors abandoned their wards to join the refugees. American health officials stepped up surveillance at U.S. airports to monitor any *Y. pestis* carriers from India (none turned up). More than a dozen foreign airlines suspended all flights to the country as cases appeared in other major Indian cities. The official death count was below a hundred out of some 5,000 reported cases, but undoubtedly many cases were missed. By the end of October, a team of American and Russian investigators for the World Health Organization declared India's cities to be plague-free but recommended that travelers avoid Surat and rural areas near Bombay, where bubonic plague had arisen. All in all, the frightening episode underlined the effect of India's paltry budget for public health, which was proportionately less than that of Bangladesh, Rwanda, Guatemala, or Pakistan. Pneumonic plague is ranked as a "Category A," or most dangerous, potential bioterrorist agent.

So don't feed the squirrels.

• Zika fever •

European explorers and colonizers of Africa who managed to survive sundry killer microbes against which they had no natural immunity nonetheless might have felt pale and sweaty, even on good days. Exotic afflictions like this one probably explain why.

An often mild or inapparent disease caused by a flavivirus (an arbovirus genus), Zika fever comes with the scenery in many parts of Africa and Asia. Half of native Africans in some regions have tested positive for antibodies, which means that they have been infected. Yellow fever, dengue, West Nile, and several varieties of hemorrhagic fever and encephalitis are also brought by flaviviruses, but Zika is evidently not as dangerous as these others to humans. The few diagnosed cases on record have shown the symptoms of fever, chills, headache, nausea, and rash so typical of tropical or subtropical insect-borne illnesses. A closely related virus causes Spondweni fever, named after the South African district. The precise cycle of infection for Zika and Spondweni is unknown but may involve livestock, which have been found to carry antibodies.

The white man perceived Africa as being "dark" for obvious reasons. Part of the great continent's mystery and threat no doubt stemmed from exotic flaviviruses like Zika.

(You'll never see it, but we need a *Z*.)

• Notes to the Introduction •

1. They are, after all, very small. A bacterium may weigh in at 0.00000000001 gram, compared to about 330,000 grams for an average adult human. There is no argument about which creature will have the longest tenure on Earth. There are somewhere between 300,000 and a million species of bacteria and about 5,000 known viruses. Only a small fraction have been scientifically characterized.

2. World Health Organization, *The World Health Report 2008,* p. 8.

3. Between 1928, when Alexander Fleming accidentally discovered that a *Penicillium* mold would kill *Staphylococcus* bacteria, and 1965, when optimism about conquering infectious disease was at its apex, more than 25,000 different antibiotics were developed. About a hundred are in common use.

4. For someone who probably spent most of his time selling buttons and ribbon, Leeuwenhoek (1632–1723) was a remarkable intellectual adventurer. His microscopes were hardly crude. The best could magnify 270 times to differentiate objects 0.0014 millimeters apart. Given that he was the first person ever to gaze upon the creepy-crawly world of microorganisms, he was also amazingly brave—he even examined his own diarrheal excrement.

5. Thus, until little more than a century ago, medical practice was based almost entirely on rule of thumb—a dangerous situation, needless to say. The highfalutin manner of the profession's establishment was perhaps akin to Dr. Marvel's bombast in *The Wizard of Oz*. See McGrew, *Encyclopedia of Medical History*, pp. 25–30.

6. Krieg, *Epidemics in the Modern World*, pp. 4, 21–22. Also, McNeill, *Plagues and Peoples*, pp. 9–14. Yellow fever epidemics in colonial Philadelphia were called "Barbados distemper" and, later, "Palatine fever," depending on the origin of recent immigrants. The epidemic of 1793 took on highly specific political overtones when pro-English Federalists held radical Frenchmen from Haiti responsible for the contagion, while liberal Jeffersonian Republicans countered that the fever had domestic origins. Both sides were right, in that the disease required local mosquitoes *and* an initial pool of infected persons. See Kraut, *Silent Travelers*, pp. 25–30. In the case of the precipitous decline of the native population in the Hawaiian Islands after Captain James Cook's arrival in 1778, a disaster of gothic intensity, foreigners' germs were entirely to blame. See A.W. Crosby, "Hawaiian Depopulation as a Model for the American Experience," in Ranger and Slack, *Epidemics and Ideas*, Chapter 8.

7. Lederberg et al., *Emerging Infections*, pp. 42–43.

8. Fee and Fox, eds., *AIDS: The Making of a Chronic Disease*, p. 26.

9. There are lots of them. An average person carries about 100 billion microbes on the outside of his body and 15 trillion inside. A normal bowel movement contains 100 trillion, since we let go of more than we keep.

10. Foucault, *Naissance de la clinique* (*The Birth of the Clinic*), p. 6. More than two millennia earlier, Aeschylus wrote that health and disease are neighbors "with a common wall."

11. The Commonwealth Fund, *Mirror, Mirror on the Wall: An International Update on the Comparative Performance of American Health Care*, vol. 59, 2007.

12. That is, medicine has always been cloistered when there was an economy to support it as such. In the United States before the

twentieth century it was a shaky and often shady occupation. See Starr, *The Social Transformation of American Medicine.*

13. Tuchman, *A Distant Mirror,* pp. 106–7.

14. Rosner, *Medicine in the Bible and the Talmud,* pp. 102–7. See also Kraut, *Silent Travelers,* p. 141. The New Testament somewhat reversed this view: "They that be whole need not a physician" (Matthew 9:12).

15. McNeill, *Plagues and Peoples,* pp. 236–37.

16. Brody, *The Healer's Power,* p. 15. "Ante Patientum Latina lingua est," advises a medieval medical maxim: "In the presence of the patient, Latin is the language."

17. In her 1979 essay "Illness as Metaphor," Susan Sontag cited Karl Menninger's observation that "the very word 'cancer' is said to kill some patients" in support of secretiveness on the part of doctors (pp. 6–7). The basic notion is very old. "A physician is only a consoler of the mind," wrote Petronius in his *Satyricon* in about A.D. 50. "Pars sanitatis velle sanari fuit," Seneca wrote around the same time: "It is part of the cure to wish to be cured." A French treatise on surgery from 1316 advises doctors to "keep up the spirits of your patient with the music of the viol and the psaltery, or by forging letters telling of the death of his enemies, or (if he be a cleric) by informing him that he has been made a bishop."

18. It is certainly true that for many centuries, charisma and clout were the best the profession had to offer. The combination is not without importance today, when more than half of all people seeing a doctor do not have a diagnosable disease and most of those who do either are incurable or will improve by themselves. See Brody, *Healer's Power,* p. 19. The American Medical Association has long opposed public access to the National Practitioner Data Bank, which holds information about some 62,000 doctors, dentists, nurses, and other professionals who have been penalized for crimes, mistakes, or incompetence.

19. To be fair, we must acknowledge that medicine is not the only profession to hold such a monopoly and profit from it. Industrial capitalism, revolutions in technology, and urbanization have

fueled a broad movement toward specialization and loss of personal autonomy. Still, patients are dependent on physicians in ways that buyers are not normally dependent on sellers. See Starr, *Social Transformation,* pp. 141–42 and 225–26.

20. My use of the term "puritan" does not imply any religious point of view, though Christian fundamentalists, whom some may see as inheritors of the puritan legacy, still seem to have more nerve than others in making their views public.

21. "Any disease that is treated as a mystery and acutely enough feared will be felt to be morally, if not literally, contagious." Sontag, *Illness as Metaphor,* p. 6.

22. Kraut, *Silent Travelers,* pp. 3, 31–49.

23. Brian Pullen, "Plagues and Perceptions of the Poor in Early Modern Italy," in Ranger and Slack, eds., *Epidemics and Ideas,* pp. 101–3.

· Selected Bibliography ·

Berkow, Robert, ed. *The Merck Manual.* 16th ed. Rahway, N.J.: Merck Research Laboratories, 1992.

Bert, Walter R., ed. *The History and Conquest of Common Diseases.* Norman, Okla.: University of Oklahoma Press, 1954.

Brody, Howard. *The Healer's Power.* New Haven, Conn.: Yale University Press, 1992.

Bumgarner, John R. *The Health of the Presidents.* Jefferson, N.C.: McFarland, 1994.

Cartwright, Frederick F. *Disease and History.* New York: Thomas Y. Crowell, 1972.

Centers for Disease Control and Prevention. *Emerging Infectious Diseases.* Atlanta, Ga.: National Center for Infectious Diseases. *Morbidity and Mortality Weekly Report.* Washington, D.C.: U.S. Department of Health and Human Services.

Crosby, Alfred W. *Ecological Imperialism.* Cambridge: Cambridge University Press, 1986.

Dale, Philip Marshall. *Medical Biographies.* Norman, Okla.: University of Oklahoma Press, 1987.

Delaporte, François. *Disease and Civilization.* Cambridge: MIT Press, 1986.

Dobell, Clifford. *Antony van Leeuwenhoek*. New York: Harcourt Brace, 1932.

Durey, Michael. *The Return of the Plague*. New York: Macmillan/Gill, 1979.

Dutton, Diana B. *Worse Than the Disease*. Cambridge: Cambridge University Press, 1988.

Eberson, Frederick. *Man Against Microbes*. New York: Ronald Press, 1963.

Evans, Alfred S., ed. *Viral Infections of Humans*. 3rd ed. New York: Plenum Medical Book Co., 1989.

Evans, Alfred S., and Harry Feldman, eds. *Bacterial Infections of Humans*. New York: Plenum Medical Book Co., 1982.

Fee, Elizabeth, and Daniel M. Fox, eds. *AIDS: The Making of a Chronic Disease*. Berkeley: University of California Press, 1992.

Foucault, Michel. *Naissance de la clinique* (*Birth of the Clinic*). Paris: Presses Universitaires de France, 1963.

————. *Power/Knowledge*. New York: Pantheon Books, 1980.

Garrett, Laurie. *The Coming Plague*. New York: Farrar, Straus and Giroux, 1994.

Gottfried, Robert S. *The Black Death*. New York: Free Press, 1983.

Gregg, Charles T. *Plague!* New York: Charles Scribner's, 1978.

Harden, Victoria A. *Rocky Mountain Spotted Fever*. Baltimore: Johns Hopkins University Press, 1990.

Hill, Justina. *Germs and the Man*. New York: G. P. Putnam's Sons, 1940.

————. *Silent Enemies*. New York: G. P. Putnam's Sons, 1942.

Hoeprich, Paul D., et al., eds. *Infectious Diseases*. 5th ed. Philadelphia: Lippincott, 1994.

Hopkins, Donal R. *Princes and Peasants*. Chicago: University of Chicago Press, 1983.

Joklik, Wolfgang K., et al., eds. *Zinsser Microbiology*. 20th ed. Norwalk, Conn.: Appleton & Lange, 1992.

Kiple, Kenneth F., ed. *The Cambridge World History of Human Disease*. Cambridge: Cambridge University Press, 1993.

Kirp, David L., and Ronald Bayer, eds. *AIDS in the Industrialized Democracies.* New Brunswick, N.J.: Rutgers University Press, 1992.

Kraut, Alan M. *Silent Travelers.* New York: Basic Books, 1994

Krieg, Joann P. *Epidemics in the Modern World.* New York: Macmillan/ Twayne, 1992.

Lederberg, Joshua, Robert Shope, and Stanley C. Oaks, eds. *Emerging Infections.* Washington, D.C.: National Academy Press, 1992.

Levine, Arnold J. *Viruses.* New York: Scientific American Library, 1992.

Longmate, Norman. *King Cholera.* London: Hamish Hamilton, 1966.

Lyons, Albert S., and R. Joseph Petrucelli. *Medicine: An Illustrated History.* New York: Harry N. Abrams, 1987.

McGrew, Roderick E. *Encyclopedia of Medical History.* New York: McGraw-Hill, 1985.

McNeill, William H. *Plagues and Peoples.* Anchor/Doubleday, 1976.

Milne, Lorus, and Margery Milne, eds. *The Audubon Society Field Guide to North American Spiders and Insects.* New York: Knopf, 1986.

Morris, R. J. *Cholera 1832.* New York: Holmes & Meier, 1976.

Morse, Stephen S., ed. *Emerging Viruses.* New York: Oxford University Press, 1993.

Paul, John R. *A History of Poliomyelitis.* New Haven, Conn.: Yale University Press, 1971.

Payer, Lynn. *Disease-Mongers.* New York: John Wiley, 1992.

Postgate, John. *Microbes and Man.* 3rd ed. Cambridge: Cambridge University Press, 1992.

Quétel, Claude. *History of Syphilis.* Baltimore: Johns Hopkins University Press, 1992.

Ranger, Terence, and Paul Slack. *Epidemics and Ideas.* Cambridge: Cambridge University Press, 1992.

Reid, Robert. *Microbes and Men.* New York: Saturday Review Press, 1975.

Rosebury, Theodor. *Life on Man.* New York: Viking Press, 1969.

Rosner, Fred. *Medicine in the Bible and the Talmud.* New York: Yeshiva University Press, 1977.

Shulman, Stanford T., et al., eds. *The Biologic and Clinical Basis of Infectious Diseases.* 4th ed. Philadelphia: W. B. Saunders, 1992.

Siegerist, Henry E. *A History of Medicine.* New York: Oxford University Press, 1951.

Singleton, Paul. *Introduction to Bacteria.* 2nd ed. New York: John Wiley, 1992.

Smith, Jane. *Patenting the Sun.* New York: Anchor/Doubleday, 1991.

Sontag, Susan. *Illness as Metaphor* and *AIDS and Its Metaphors.* New York: Anchor/Doubleday, 1989.

Starr, Paul. *The Social Transformation of American Medicine.* New York: Basic Books, 1982.

Tuchman, Barbara W. *A Distant Mirror.* New York: Ballantine Books, 1979.

Twigg, Graham. *The Black Death.* London: B.T. Batsford, 1984.

Waitzkin, Howard. *The Politics of Medical Encounters.* New Haven, Conn.: Yale University Press, 1991.

Washington Post, Health.

Williman, Daniel, ed. *The Black Death.* Binghamton, N.Y: State University of New York, 1982.

Wilson, Mary E. *A World Guide to Infections.* New York: Oxford University Press, 1991.

Ziegler, Philip. *The Black Death.* Harmondsworth: Pelican Books, 1970.

Zinsser, Hans. *Rats, Lice and History.* New York: Blue Ribbon Books, 1935.

· *Index* ·

Abortion, 34
Acquired immunodeficiency syn-
    drome. *See* AIDS
"Acute respiratory disease" (ARD), 21
Acyclovir, 83
Adams, John Quincy, 146
Adenovirus, 21–22
Aedes aegypti mosquito, 38, 54, 59,
    180, 181, 182–84
Aedes albopictus mosquito, 60
Aedes triseriatus mosquito, 54
Aeschylus, 194
"Agues," 120
AIDS (acquired immunodeficiency
    syndrome), 8–9, *9,* 16, 37, 47, 58,
    81, 83–90, *85, 86,* 110, 141, 168,
    194
Amapari arenavirus, 27
American Medical Association, 195
American Psychiatric Association, 14
Amoebic disease, 64
Ancylostoma duodenale worm, 178
Anemia, 87
    See also Sickle-cell anemia

A/New Jersey/8/76 influenza, *93*
Animal bites, vs. human bites, 28–29
Animalcules, 5
Anopheles mosquitoes, 38, 120, 121
Anthrax (*Bacillus anthracis*), 23–26, *23,*
    165
Antibiotics, 4, 8, 33, 40, 47, 52–53,
    63, 64, 66, 74, 75, 78, 98, 112, 113,
    134, 136, 137, 139, 150–51, 153,
    155, 159, 164, 172, 173, 185, 190,
    193
Anus, venereal warts on, 116
Apollo (mythological character), 157
Appendicitis, 179
Arboviruses, 26–27, 37–38, 53, 191
ARD. *See* "Acute respiratory disease"
Arenaviruses, 27–28
Argentinian hemorrhagic fever, 27
Armadillos, 170
Arteries, hardening of, 40, 58
Arthritis, 87
"Arthropod-borne virus," 26–27
Ascaris lumbricoides. See Roundworm
Asian flu, 92

Asthma, 134
Astruc, Jean, 82
Atherosclerosis, 40, 58
Athlete's foot. *See* Tinea pedis
Automobile accidents, 52

B. abortus, 34–35
    See also Brucella
Bacillary dysentery, 64
Bacillus anthracis. See Anthrax
Baker's and brewer's yeast
    (*Saccharomyces cerevisiae*), 37
Bangkok (A) flu, 93
Bann, Rahima, 149
"Barbados distemper," 194
Barthélemy, Auguste Marseille, *11*
Bartholin, Thomas, *13*
Bashunto, 149
Bats, 27, 74, 131, 176
Baudelaire, Charles, 10, 160
Bayou virus, 77
B. canis, 34
    See also Brucella
Bear, 163
Beaver, 74
Beef, 51, 66, 99–100, 132, 140
Beef tapeworm (*Taenia saginata*), 177
Bell, Sir Charles, 51
Bioterrorism, 1, 10, 24, 142, 149,
    150
Birds, 26, 40–41, 74, 95–96, 129, 175
Birth defects, 58, 107, 113
Bites, 26, 28–29
Black Creek Canal virus, 77
Black Death, 13, 16–17, 83, 169, 181,
    184–85, 187, 189
Blackflies, 177
"Black measles," 136–37
Black typhus (*el typho negro*), 27
"Black-water fevers," 120
Bleeding, as medical treatment, 122

Blindness, 39–40, 113, 177
Blood poisoning, 48, 100
Blood transfusions, 80
    See also Transfusions
Blue jays, 175
B. melitensis, 34–35
    See also Brucella
Boccaccio, Giovanni, 188
Body lice (*Pediculus humanus;*
    "cootie"), 129, 135, 172, 173
Boer War, 170
Boils, 151
Bone disorders, 110
Bone lesions, 107
Bordet, Jules, 30
Bordetella pertussis (whooping cough),
    9, 29–30
Borgia, Pope Alexander, 160
Borrelia burgdorferi (Lyme disease),
    30–33, 67
"Borreliosis neurosis," 30–31
Botulism, 49–50
Bovine spongiform encephalopathy
    (BSE), 98–100
Brain abscess, 64
Brain damage, 30, 109
B. rangiferi, 34
    See also Brucella
"Break-bone fever," 59
Breast milk, 65, 80
Bretonneau, Pierre-Fidèle, 61
Bronchiolitis, 117
Bronchitis, 117
Brontë, Emily, 167
Bruce, David, 34
Brucella, 34–35
Brucellosis, 34–35
B. suis, 34
    See also Brucella
Bubonic plague, 172, 185, *186,*
    189–90

Buffalopox, 142
Bunyamwera tropical fever, 26
Bunyaviridae family, 54
Burgdorfer, Willy, 32
Burgers, 140
Burkitt's lymphoma, 73
Burnet, F. M., 129
Burns, 152
Bush, George, Sr., 153

Caesarian sections, 14
Calves, 35
California arbovirus, 53, 54
California encephalitis, 54
Calomys genus, 27
Camelpox, 142
Campylobacter, 35–36
Cancer, 22, 73, 80, 82, 86, 162
    cervical, 116
    liver, 82
    throat, 73
Candida albicans, 36–37
Candidal jamborees, 37
Candida utilis. See Food yeast à la
    Marmite
Candiru tropical fever, 26
Caribou, 34
Cataracts, 107
Catherine the Great, empress of
    Russia, 146
Cats, 87, 131, 176, 185
    See also Kittens
Cattle, 23, 34–35, 36, 48, 53–54, 56,
    66, 71, 98, 99–100, 104, 129, 164
CD4 lymphocyte, 87
Cellini, Benvenuto, 160
Centers for Disease Control and Pre-
    vention, 25, 86, 88–89, *102,* 149
    Epidemic Intelligence Service
        of, 76
Cerveau, 100

Cervical cancer, 116
Cestodes, 176
CFIs. *See* Clenched-Fist Injuries
C. glutamicum, 63
    See also Corynebacterium diptheriae
Charlemagne, Frankish king and
    emperor of the West, 91, *189*
Charles VIII, king of France, 159
Chaude pisse (hot piss), 111
Cheese, 34
Chekhov, Anton, 167
Chest X rays, 166
Chicken feed, 140
Chicken pox, 9, 22, 142, 174–75
Chickens, 92, 95, 140
Chiggers, 129, 135
Chikungunya, 26–27, 37–38
"Chin cough," 29–30
Chinese mushrooms, 152
Chinese restaurant syndrome, 38
Chimpanzees, 117, 126
Chlamydia, 39–41
Chlorination, 67, 69, 97
Chloroquine, 124
Cholera (*Vibrio cholerae*), 35, 41–48,
    43, 44, 45, 163, 165
Cholera morbus, 46
Chopin, Frédéric, 167
"Chronic fatigue syndrome," 72–73
Cinchona (quinaquina), 122, 123
Cirrhosis, 81
Civil War, 21, 64, 79, 121
Clams, 70
"Clap," 111
Clenched-Fist Injuries (CFIs),
    28–29
Clostridium, 48–53
Clostridium botulinum, 49–50
Clostridium difficile, 52–53
Clostridium tetani, 51–52, 62
Cocaine, crack, 161

Coccidioides immitis (fungus), 141
Coccidioidomycosis, progressive, 141
Colds, 55, 69, 90, 92, 133–34
Cold sores, 82–83
Colitis, 52
   hemorrhagic, 66
Colorado tick fever, 54–55
Columbia, 183
Columbus, Christopher, 159, 173, 180
Condoms, 40, 81, 114, 116, 161
Congo arbovirus, 53
Conjunctivitis, 40, 69, 93
"Consumption." *See* Tuberculosis
Convulsions, 23
Cook, James, 194
"Cootie." *See* Body lice
Copernicus, Nicolaus, 157
Coqueluche, 29–30
Cordons sanitaires, and cholera, 42
Coronary bypass operations, 14
Coronary heart disease, 39
Coronaviruses, 55–56
Cortés, Hernando, 104
Corynebacterium diphtheriae. See
   Diphtheria
Corynebacterium glutamicum, 38
Cotton rat, 77
Coughs, 91
Cowpox, 142
Cows, 34, 87, 98, 147, 165
Cox, H. R., 129
Coxiella burnetii, 124
Coxsackie viruses, 69, 71
C. pasteurianum, 53
   See also Clostridium
C. perfringens, 51
   See also Clostridium
C. pneumoniae, 39, 40
   See also Chlamydia
C. psittacii, 39, 40
   See also Chlamydia

Crabs, 44–45
Crack cocaine, 161
Cramps, 46, 51, 56, 75
Creutzfeldt-Jakob disease, 98–99
Crimean arbovirus, 53
Cromwell, Oliver, 122
Croup, 117
Crows, 175
Cryptosporidiosis, 56
Cryptosporidium, 56–57
Cryptosporidium parvum, 56
C. trachomatis, 39
   See also Chlamydia
Culex mosquito, 54
Cutaneous anthrax, 23–24, 25
Cytomegaloviruses, 57–58

Dalrymple, David (quoted), 122
"Dandy fever," 59
Darwin, Charles, 36, 132–33
DDT, 121, 173, 188
Deafness, 107
Deer, 31, 33, 67, 71, 74
Deer mice (*Peromyscus maniculatus*), 76
Deer ticks, 32–33, 136
Delousing, 173
Democritus, 130
Dengue fever, 26, 37–38, 58–60, 191
Dengue hemorrhagic fever, 60
Department of Agriculture, 66
Dermacentor andersoni (wood tick),
   54–55
Desert rheumatism, 141
Diabetes
   insulin-dependent, 71
Diaper rash, 37
Diarrhea, 35, 51, 53, 56, 65, 68–69,
   75, 100, 103, 138, 139, 193
Diphtheria (*Corynebacterium
   diphtheriae*), 30, 38, 61–63
"Discharge," 111

Disinfectants, 69
DNA, 87, 149
Dogs, 28, 34, 67, 74, 104, 130–31,
  132
  See also Prairie dogs; Puppy
Dog ticks, 136
Domestic rat (Rattus Rattus), 185
Dracunculus medinensis. See Guinea
  worm disease
Drainage systems, 69
Drug abuse, 81
Dryden, John, 79
DTP shot, 30, 62
Ducks, 80, 93
Duke's disease, 141
"Dumb rabies," 131
Dysentery, 63–64, 66

Earache, 140
Ear infections, 155
Eastern equine encephalitis, 54, 60
Ebola virus, 100–3, 101, 102
Echoviruses, 69
E. coli, 65–67
E. coli 0157:H, 66–67
  See also Escherichia coli
Ectopic pregnancy, 40
Edward I, king of England, 63
E. ewingii, 67
E. faecalis, 68
  See also Enterococcus
Eggs, 139, 140
"Egyptian" ophthalmia, 40–41
Ehrlich, Paul, 160
Ehrlichia chaffeensis, 67, 135
Ehrlichiosis, 65
Elephantiasis, 177
Elk, 67
Encephalitis, 26, 53, 54, 68–69,
  175–76, 191
  eastern equine, 54, 60

Endemic typhus. See Murine typhus
"Energy parasites," 39
Entamoeba histolytica, 64
Enteric pathogens, 35–36
Enterobius vermicularis. See Pinworm
Enterococcus, 68
Enterococcus faecium, 68
Enteroviruses, 68–71
Environment. See "Miasma" theory of
  disease
E. phagocytophila, 67
Epidemic Intelligence Service, of
  Centers for Disease Control, 76
Epidemic typhus (jail fever), 172–73
Epstein-Barr virus, 71–73
Erythritol, 35
Escherichia coli. See E. coli
E. sennetsu, 67–68
Evans, Alice, 35
Excreta, 27
Eye infections, 39

Fast-food hamburgers, 65, 66
Feces, 74–75, 79, 81, 126, 135, 138,
  170, 177–78
  See also Excreta
Female genital tract problems, 179
Fever
  blisters, 82–83
  dengue hemorrhagic, 60
  hemorrhagic, 27, 53, 191
  undulant, 35
Fifth disease. See Slapped cheek
  disease
Filariasis, 177
Filoviruses, 100
Finlay, Dr. Carlos, 183–84
Fish, 74
Flamingos, 175
Flaubert, Gustave, 160
Flavivirus, 191

Fleas, 172
  Oriental rat, 185
  rat, 135
  water, 177
Fleming, Sir Alexander, 133, 150–51, 193
"Flesh-eating" bacterium, 156, 157
Flies, 26, 40, 129
Flu, 21, 78, 88, 100, 108, 110, 124, 138, 152, 176
  See also Influenza
Food poisoning, 49, 51, 98, 139, 152
Food yeast à la Marmite (*Candida utilis*), 37
Foot-and-mouth disease (FMD), 71
Ford, Gerald, *94*, 94
Forest canopy mosquitoes, 184
Fort Bragg diseases, 73–74
Foucault, Michel, 10, 194
Foxes, 74, 131–32
Fox squirrels (*Sciurus niger*), 189
Fracastoro, Girolamo, 157
Franklin, Benjamin, 146
"French disease," 157
  See also Syphilis
Frogs, 74
Fungi, 4
"Furious rabies," 131

Galen, 166
Gametocytes, 121
Gangrene, 156
García Márquez, Gabriel, 41
Gastroenteritis, 138, 139
Gastrointestinal anthrax, 25
Gastrointestinal bacteria, 140
Genitalia, venereal warts on, 116
Genital infections, 39–40
German measles, 107
  See also Rubella
Germ theory of disease, 5–6

Giard, Alfred M., 75
Giardia lamblia, 74–75
Gibraltar fever, 34
Gnats, 53
Goats, 23, 24, 34, 53, 74, 99
Gokwomyo, Emperor, 143
Gonorrhea, *20,* 39, 110, *158,* 160
  See also Neisseria gonorrhoeae
Gout, 160
Grant, Ulysses S., 121
Great Fire of 1666, 188
Gregg, Norman, 107
"Grippe, la," 91
Guillain-Barré syndrome, 94
Guinea pig, 96, 126
Guinea worm disease (*Dracunculus medinensis*), 177
Gulf War, 50
Gynecological problems, 112–13

Ham, 49
Hamburgers, fast-food, 65, 66
Hand, foot, and mouth disease (HFMD), 71
Hansen, Armauer, 169
Hansen's disease, 169
Hantaan virus, 75, 76
Hantaviruses, 75–77
Hantavirus pulmonary syndrome (HPS), 75–76
Hard-backed ticks. *See* Ixodidae ticks
Hawks, 175
Hearst, William Randolph, 171
Heart attack, 40, 58
Heart disease, 86, 154–55
  See also Coronary heart disease
Heart problems, 90, 107, 109
Hedgehogs, 53
Heine, Heinrich, 160
Helicobacter pylori, 77–78
Hemolytic uremic syndrome, 66

Hemophiliacs, 86
Hemophilus, 78
Hemorrhages, 100
Hemorrhagic colitis, 66
Hemorrhagic fever, 27, 53, 191
Hemorrhoids, 63–64
Henry V, king of England, 63
Henson, Jim, 156
Hepatitis, 78–81, 88
    A, 57, 70, 78–80
    B, 78–79, 80–81
    C, 78–79, 81
    D, 78–79, 81
    E, 78–79, 81
Herodotus (quoted), 82
Herpes, 22, 58, 72, 82–83, 174
Herpes gladiatorum, 83
Herpes simplex 1 (HSV-1), 82–83
Herpes simplex 2 (HSV-2), 82–83,
    116
Herpes zoster. See Shingles
H. influenzae type b. (Hib), 78
    See also Hemophilus
Hippocrates, 51, 63, 119, 164
Hippocratic oath, 12
HIV (human immunodeficiency
    virus), 83–90, 116
Hoagland, R. J., 71
Hogs, 34
Home canning, 49
Homosexuality, 14
Honey, 50
Hong Kong flu, 92
Hookworm disease, 178
Horses, 23, 34, 36, 48, 52, 175, 176
Horus (mythological character), 63
Hot dogs, 98
Hot piss. See Chaude pisse
HSV-1. See Herpes simplex 1
HSV-2. See Herpes simplex 2
Human bites, vs. animal bites, 28–29

Human granulocytic ehrlichiosis
    (HGE), 67
Human immunodeficiency virus.
    See HIV
Human inhalation anthrax, 25–26
Huxham, John, 91
"Hydrophobia," 131

Ice cream, 140
I. dammini, 32
Immune system, 22, 36, 48, 52, 56,
    58, 65, 71, 76, 82, 83, 87–88, 89,
    91, 127, 139, 141, 149, 152, 168,
    174
Impotence, 100
"Infantile paralysis." See Polio
Infections, 6–9
Infectious hepatitis, 79
"Infectious jaundice," 74
Infertility, 40, 112
Influenza, 30, 90–96, 155
    A, 93, 94, 95
    A/New Jersey/8/76, 94
    C, 93
    See also Flu
Insecticides, 121, 122
Insects, 26, 53, 59, 121, 134–35, 176,
    191
Insulin-dependent diabetes, 71
Intercourse, sexual, 158
Intestinal illnesses, 35
I. pacificus ticks, 32
I. scapularis ticks, 32
Ivan the Terrible, 160
Ixodidae (hard-backed) ticks, 67, 136

Jackson, Andrew, 120
Jail fever. See Epidemic typhus
Jaundice, 72, 74, 78
Jefferson, Thomas, 148, 181
Jenner, Edward, 106, 147–48, 148

"Jock itch." *See* Tinea cruris
"Jolly rant." *See* Influenza
Joseph I, Emperor of Austria, 142
Junin virus, 27
Justinian, Emperor, 186

Kant, Immanuel, 146
Kaposi's sarcoma, 89
Keats, John, 167
Kidney disease, 66, 74
Kipling, Rudyard, 12
Kissing, 71, 134, 158
Kitasato, Shibasaburo, 52
Kittens, 36, 132
Klappor. See Neisseria gonorrhoeae
Koch, Robert, 6, 7, 24, 42–43, 48,
    165–66
Kōmei, Emperor, 143
Koop, C. Everett, 86
Korean hemorrhagic fever, 75–76
Korean War, 115
Kyasanur tropical fever, 26

Laënnec, René-Théophile-Hyacinthe,
    164
Lassa fever, 27, 101
Latino arenavirus, 27
Lawrence, D. H., 167
Lederberg, 86
Leeuwenhoek, Antoni van, 5, 74–75,
    193
Legionella ("Legionnaire's disease"),
    96–97
Legionellosis, 97
Lentiviruses, 87, 88
Leonardo da Vinci, 14
Leprosy, 110, 159, 169
Leptospira, 73–74
Leptospirosis, 74
Lice. *See* Body lice
Lincoln, Abraham, 149

Lindane (pesticide), 173
Listeria monocytogenes, 98
Liver cancer, 80
Liver damage, 64, 74, 100, *101*
Liver disease, 48, 57, 78, 81
Lizards, 32
Lobar pneumonia, 124
Locke, John, 164
"Lockjaw," 51
Lone star tick, 135
Louis XIV, king of France, 122
Louis XV, king of France,
    142–43
Lower respiratory disease, 117–18
L. pneumophila, 97
    See also Legionella
Lung disease, 90
Lyme disease. *See Borrelia burgdorferi*
Lymphogranuloma venereum, 39
Lymphoma, Burkitt, 73
Lymph-system problems, 110

Maalin, Ali Maow, 149
Machupo arenavirus, 27
Mad cow, 98–99
M. africanum, 164
Malaria, 38, 59, 118–24, *119, 120,*
    123, 182, 184
    See also Plasmodium
Malathion (pesticide), 173
Mallards, 175
Malnutrition, 103, 168
Malone, Mary (Typhoid Mary), 170,
    171
Malta fever, 34
Mann, Thomas, 160, 167
Marburg virus, 100–3
Marijuana, 139
Marituba tropical fever, 26
Marmots, 185
Mary II, queen of England, 142

Mather, Cotton, 61, 146, 181
Maupassant, Guy de, 160
Mayaro tropical fever, 26
M. bovis, 164
Measles, 9, 88, 103–7, *105,* 110, 117,
   124, 141, 154, 168
   See also "Black measles"
Meat, 65, 66, 132–33, 163, 177
Meister, Joseph, 130
Meningitis, 27, 68, 70, 74, 78, 98,
   114–15, 135
Meningococcal disease, 115
Menninger, Karl, 195
Mercury, 160
Merozoites, 121
"Miasma" (environmental) theory of
   disease, 120, 182
Mice, 27, 32, 33, 52, 70, 74, 126,
   185
   deer, 76
   white-footed, 32, 77
Microsporum, 162
Milk, 34, 66, 129, 140, 154, 172
   See also Breast milk
Mites, 26
MMR shots, 108, 110
Monkeypox, 142
Monkeys, 70, 88, 100–1, 102, 126,
   127, 184
Mononucleosis, 57, 71–72
Monosodium glutamate. *See* MSG
Montagnier, Luc, 8
Montagu, Lady Mary Wortley, 145,
   145
"Montezuma's revenge," 65
Mooseri. *See Rickettsia typhi*
Mosquitoes, 26, 89, 119–20, 121–23,
   176, 182, *183,* 194
   Aedes aegypti, 38, 54, 59–60, *180,*
      181–82
   Aedes albopictus, 60

Aedes triseriatus, 54
Anopheles, 37, 119–20, 121
Culex, 54
forest canopy, 184
Mouse mites, 135
Mozart, Wolfgang Amadeus, 146
MSG (monosodium glutamate), 38, 63
Muerta Canyon virus, 76
Mumps, 9, 22, 103, 108–10, *108,*
   109, 117, 168
Murine (endemic) typhus, 172
Muscle weakness, 49
Mushrooms, Chinese, 152
Mussels, 70
Mycobacteria, 110
Mycobacterium leprae, 169
Mycobacterium tuberculosis, 164,
   165–66

Nader, Ralph, 94
Nairoviruses, 53
Napoleon, 63, 120, 173, 182
National Practitioner Data Bank,
   195
Necatur americanus, 178
"Necrotizing fasciitis," 156–57
Neisser, Albert, 113
Neisseria gonorrhoeae (gonorrhea),
   111–14, *111, 112*
Neisseria meningitidis, 114–15
   A, 115
   B, 115
   C, 115
Nematodes, 176
Neonatal tetanus, 52
Neosalvarsan, 160
Nerve disorders, 110
"New acquaintance, the" 91
New Jersey (A) flu, *93*
New York-1 virus, 77
Nightingale, Florence, 41–42

Nitrites, 50
Nixon, Richard, 25, 35
Norwalk (Ohio) viruses, 138
Nose, cancer of, 73

Ogston, Sir Alexander, 151
Onchocerra volvulus, 177
O'nyong-nyong virus, 37–38
Ophthalmia, "Egyptian"
   ("military"), 40
Opium, 14
Oriental rat flea (*Xenopsylla cheopsis*),
   185
Orthopoxvirus, 142, 149
Orwell, George, 167
Oysters, 44–45, 48, 70

Pancreas, complications in, 109
Pandemics, of cholera, 41–44
Papillomaviruses, 116
Parainfluenza virus, 116–18
Paralytic poliomyelitis, 70, 125, 127
Paramyxovirus, 103, 108, 117
Parana arenavirus, 27
Parotid (salivary) glands, swelling of,
   108
Parvovirus B19, 141
Pasteur, Louis, 5–6, *5,* 24, 124, 130
Pasteurization, 35, 129, 140, 154, 165,
   172
Pasteurella, 28
Pathogens, 4–8
   enteric, 35–36
Pediculus humanus. See Body lice
Peloponnesian War, 63
Pelvic inflammatory disease, 40, 112
Penicillin, 8, 28, 113, 125, 151, 156,
   159, 161, 190
Penicillinase-producing *Neisseria
   Gonorrhoeae* (PPNG), 113
Penicillium mold, 150–51, 193

Pepys, Samuel, 154, 188
Peromyscus leucopus. See White-footed
   mice
Peromyscus maniculatus. See Deer mice
Pertussis (whooping cough), 62
Pesticides, 173
Peter the Great, czar of Russia, 160
Petronius, 195
P. falciparum, 121
   See also Plasmodium
Pharyngitis, 72
Phipps, James, 147
"Phthisis." *See* Tuberculosis
Pichinde arenavirus, 27
Pigeons, 41
Pigs, 23, 74, 92–93, 163
Pill, the, 113
Pinworm (*Enterobius vermicularis*),
   178–79
Placebo effect, 15
Plague, *13,* 16, 91, 180, 184–91, *185,*
   187, 190
   bubonic, 172, 185, *186,* 189–90
   pneumonic, 190, 191
"Plague Pit, The," *187*
Plasmodium, 118–24, 120
   See also Malaria
Pliny the Younger, 133
Pneumococcus, 124–25, 155, *155*
Pneumonia, 21, 30, 39, 40, 90, 95, 96,
   103, 117, 124–25, 137, 152, 155,
   171
pneumocystis carinii, 88–9
Pneumonic plague, 190, 191
Pocahontas, 142
Polio (poliomyelitis), 68–69, 70, 79,
   90, 114, 125–28, *125, 128*
Polioviruses, 69, 126–28, 133
Pontiac fever, 96
Pork, 163
Pork tapeworm (*Taenia solium*), 177

Porter, Katherine Anne, 92
Potato salad, 98
Potemkin, Grigori, 173
Poultry, 36, 51
PPNG. *See* Penicillinase-producing
  Neisseria gonorrhoeae
Prairie dogs, 185
Pregnancy, 58, 141
  ectopic, 40
Presley, Elvis, *128*
Procopius, 186
Progressive coccidoidomycosis, 141
Protozoa, 4
Prowazek, Stanislas von, 172
Prusiner, Stanley B., 99
Psittacosis, 40
Psychosis, 100
Pulmonary anthrax, 24
Punta Toro tropical fever, 26
Puppy, 36
Puulmala virus, 76

Q fever, 129–30
Quarantine, *18*
Quinaquina. *See* Cinchona
Quinine, 122–23
Quinta. See Bordetella pertussis

Rabbitpox, 142
Rabbits, 53, 74, 130, 176, 185
Rabies, 74, 119, 130–33, *132,* 133
  wolf, 130
Raccoonpox, 142
Raccoons, 71, 131–32
Ragpicker's disease, 24
R. akari, 150
  See also Rickettsia
Ramses V, king of Egypt, 142
Rat fleas, 135
Rats, 27, 75, 186, *186,* 187–88
  See also Domestic rat

Rattus Rattus. See Domestic rat
Reagan, Ronald, 86
Rectum, infection of, 113
Reed, Dr. Walter, 184
Reindeer, 34
Respiratory illnesses, 21, 49, 55, 57,
  68–69, 103
  See also Lower respiratory disease;
  Upper respiratory problems
Respiratory syncytial virus (RSV),
  116–18
"Reston" virus, 102
Retroviruses, 87
Reverse Transcriptase, 87
Revolutionary War, 146
Rhazes, al- (Abu Bakr Muhammad
  ibn Zakariyā), 103–4
Rheumatic fever, 154–55
Rheumatism, desert, 141
Rhinoviruses, 55, 133–34
Ricketts, Howard T., 134, 137,
  172
Rickettsia, 134–38, 172
Rickettsia burnetti, 129
Rickettsialpox, 134
Rickettsia prowazekii, 134, 172
Rickettsia rickettsii, 134, 136, 137
Rickettsia typhi (mooseri), 172
Rift Valley fever, 54
Ringworm. *See* Tinea
Risus sardonicus, 57
River blindness, 177
RNA, 87
Rockefeller Foundation, 184
Rocky Mountain spotted fever, 67,
  129, 135, 136–38, 173
Rodents, 28, 67, 76, 132, 135, 189
Roosevelt, Franklin D., 70, 127
Roseola, 141
Rossalia. See Scarlet fever
Rotaviruses, 138

Roundworm (*Ascaris lumbricoides*),
177, *178*
R. *quintana,* 135
See also Rickettsia
RSV. *See* Respiratory syncytial virus
R. *tsutsugamushi,* 135
See also Rickettsia
Rubella (German measles), 22, 58,
107, 108, 141
Rush, Dr. Benjamin, 59, 182

Sabin, Albert, 94
Sabin vaccine, 125
Saccharomyces cerevisiae. See Baker's and
brewer's yeast
S. *agona,* 140
See also Salmonella
S. *albus,* 151
See also Staphylococcus
Salivary glands. *See* Parotid glands
"Salivary-gland virus," 58
Salk, Jonas, 127
Salk vaccine, 125, 126
Salmon, Daniel E., 139
Salmonella, 35–36, 139–40
Salmonella typhi, 165
Salvarsan, 160
Sandflies, 26
San Joaquin fever, 141
Sarsaparilla, 160
Sauerkraut, 134
S. *aureus,* 151, 152
See also Staphylococcus
Sausage, 163
Scarlet fever, 141, 153
Schistosoma, 178
Sciurus niger. See Fox squirrels
Scrapie, 99
Scrofula, 164
Scrub typhus, 135
Seafood, 45, 48

Seagulls, 93
Seneca, 195
Seoul virus, 76
S. *epidermidis,* 151
See also Staphylococcus
Serum hepatitis, 79
Sewage systems, 69
Sexual abuse, 113
Sexually transmitted diseases, 116,
160–61
Shakespeare, William, 82, 130, 160
Sheep, 23, 34, 71, 99, 129
S. *heidelberg,* 139
See also Salmonella
Shelley, Percy Bysshe, 167
Shellfish, 70, 79
Shigella, 35
Shigella dysenteriae, 64
Shigella flexneri, 64
Shigella sonnei, 64
Shigu, Kiyoshi, 64
Shingles (herpes zoster), 174
Shitala Mara, Indian goddess of
smallpox, *144*
Shrimp, 48
Sickle-cell anemia, 122, 141
Sin Nombre (no-name virus), 76
Skin herpes, 83
Skin-prick test, for tuberculosis,
166
Skin ulcers, 110
Skin wart, 116
Skunks, 131–32
Slapped cheek disease, 141
Smallpox, 9–10, 104, 106, 126,
142–50, *143, 144, 147, 148,* 163,
174
S. *montevideo,* 140
See also Salmonella
Snails, 178
Snakes, 28, 74

S. newport, 140
　See also Salmonella
Snow, John, 41–42
Sontag, Susan, 195
Soper, George, 171
Sopona, god of smallpox, *143, 144*
Sores, cold, 82
Sore throats, 153, 154, 157
Spanish-American War, 170
"Spanish disease." *See* Syphilis
"Spanish flu," 91–92
Spiders, 98
Spinal cord, inflammation of, 125
Spleen, swelling of, 72
S. pneumoniae, 124, 125
　See also Pneumococcus
Spondweni fever, 191
Sporozoites, 121
Spotted fever, 137
Squirrels, 80, 132, 176, 185, 188, 191
S. saint paul, 139–40
　See also Salmonella
Staphylococcus, 28, 66, 150–53, *152,*
　193
Stevenson, Robert Louis, 167
St. Louis arbovirus, 54
St. Louis encephalitis, 54, 175
"Stomach flu," 91
Stomach ulcers, 77–78
Strain, Earl, 137
Strangouria (urinary blockage), 111
Strep-A bacteria, 156
Strep-B strains, 156
Streptococcus, 68, 152–53, *154*
Streptococcus pneumoniae, 124, 157
Streptomycin, 166
Stress, 134
S. typhi, 170–71
　See also Typhoid
Sulfa drugs, 113
Swine flu, 92, *94,* 94–95

Sydenham, Thomas, 144
Syphilis (*Treponema pallidum*), 6, 10,
　11, 20, 87, 142, 157–62, *158*

Tacaribe arenavirus, 27
Taenia saginata. See Beef tapeworm
Taenia solium. See Pork tapeworm
Tamiami arenavirus, 27
Taylor, Zachary, 173
TB. *See* Tuberculosis
T cell, 87
Tennyson, Lord Alfred, 71
Testicles, swelling of, 108
Tetanus, 30, 51–52, 62
Tetracycline, 47, 113, 190
Thirty Years' War, 173
Thoreau, Henry David, 167
Throat
　cancer of, 73
　infections of, 113
Thrush, 37
Thyroid, complications in, 109
Tick-borne typhus, 130
Tick fevers, 131
Ticks, 26, 53, 89, 129, 135, 136,
　137–38
　deer, 32–33, 136
　dog, 136
　I. pacificus, 32, 67
　I. scapularis, 32–33, 67
　Ixodidae, 67, 136
　wood, 54–55, 136
Timoni, Emanuel, 145
Tinea (ringworm), 162
Tinea barbae, 162
Tinea capitis, 162
Tinea cruris ("jock itch"), 162
Tinea pedis (athlete's foot), 162
T lymphocytes, 87
T. microdentium, 162
　See also Treponema pallidum

Tonsillectomy, 153
Toulouse-Lautrec, Henri de, 160
Toxic shock syndrome, 152
Toxoid, 52
T. pertenue, 161
    See also Treponema pallidum
Tracheotomies, 61
Trachoma, 39
Transfusions, 89
    See also Blood transfusions
Transmissible spongiform
    encephalopathy (TSE), 99
Traveler's tummy, 65
Trematodes, 176
Trench fever, 135
Treponema pallidum. See Syphilis
Trichinella spiralis, 163
Trichinosis, 163
Trichophyton infection, 162
Trojan War, 186
Tropical fever malaises, 26
"Tuberculin," 166
Tuberculosis (TB), 110, 163–69, *165,
    166, 167*
Tumors, 22
Turtles, 139
"Tuskegee experiment," 161
Typhoid, 64, 101, 170–72
Typhoid Mary. *See* Malone, Mary
Typho negro, el. See Black typhus
Typhus, 129, 134, 132, 165, 167–69
    scrub, 135
    "tick-borne," 134–35
    See also Black typhus

Ulcers
    skin, 110
    stomach, 77–78
Ultraviolet rays, 168
Undulant fever, 35
UNICEF, 47

Upper respiratory problems, 63, 117
Urinary blockage. *See Strangouria*
Urinary tract infections, 40, 110
Urine, 73, 178
U.S. Public Health Service, 161

V. alginolyticus, 48
    See also Cholera
Vampires, 29
Van Ermengem, 49
Varicella-zoster virus, 174–75
Variola major, wild, 149
Variola minor, 149
Varro, 5
V. cholerae 01, biotype El Tor, 43
    See also Cholera
V. cholerae 0139, 44
    See also Cholera
Venereal disease, 80, 87, 111, 113,
    157
Venereal warts, 116
Verdi, Giuseppe, 167
Vibrio cholerae. See Cholera
Vibrio fetus, 36
Vibrio germs, 35
Vietnam war, 115
Viral encephalitis, 59
Viruses, 4–5, 9
Virus Watch Programs, 69
Vitamin C, 134
Voles, 75
Voltaire, 146
V. parahemolyticus,
    See Cholera
V. vulnificus,
    See Cholera

Walrus meat, 163
Wart, 116
Washington, George, 61, 63–64, 120,
    146, 147

Water, 41, 46–48, 56–57, 68, 69–70, 73, 74, 75, 79, 97, 98, 138, 170, 172, 177
Water fleas, 177
Waterfowl, 93
Weeksiella, 28
Weil, Adolph, 74
Weil's syndrome, 74
West Nile virus, 26, 175–76
Western equine encephalitis, 54
White-blood-cell count, low, 135
White-footed mice (*Peromyscus leucopus*), 32, 77
Whitewater Arroyo, 28
Whitman, Walt, 167
Whooping cough. *See Bordetella pertussis*
"Wild" polioviruses, 127
William the Conqueror, 63
Williams, William Carlos, 61
Willowbrook State School for the Mentally Handicapped, 79
Wizard of Oz, The (film), 194
Wolf rabies, 130
Woodchucks, 80
Woodrats, 28
Wood ticks, 54–55, 136
Woolsorter's disease, 24
World Health Organization, 84, 123, 127, 149–50, 190

World War I, 52, 74, 109–10, 115, 124, 135, 155, 160, 170, 173
World War II, 21, 37, 53, 79, 110, 113, 115, 121, 123, 129, 151, 155, 162, 173, 188
Worms, 176–79
Worry wrinkles, 50
Wound infections, 48–50
Wünder-Bullet ("magic bullet"), 160

Xenopsylla cheopsis. See Oriental rat flea
X rays, chest, 166

Yaws, 161
Yeast infections, 36
Yellow fever, 26, 38, 59, 81, 101, 126, 180–84, 191
Yellow Fever Commission, 184
"Yellow Jack," 182, *183*
Yeltsin, Boris, 25
Yersin, Alexandre, 188
Yersinia pestis, 167, 184–91
Yogurt, 154
Yorubas, the, 143
"Yuppie disease," 72–73

Zakariyā, Abu Bakr Muhammad ibn. *See* Rhazes, al-
Zika fever, 191
Zoonotic disease, 23, 34

• *Acknowledgments* •

I wish to thank the staffs of the National Library of Medicine in Bethesda, Maryland; the Johns Hopkins University libraries in Baltimore; and the Library of Congress and George Washington University library in Washington, D.C., for their generosity.